Safety and Security in Building Design

Safety and Security in Building Design

Ralph Sinnott

MSc, PhD, FCIOB

VNR VAN NOSTRAND REINHOLD COMPANY
———— New York ————

First published in Great Britain by
Collins Professional and Technical Books 1985

Copyright © by Ralph Sinnott 1985

ISBN 0—442—28212—5

Printed and bound in Great Britain

Published by Van Nostrand Reinhold Co. Inc.
135 West 50th Street
New York, NY 10020, USA

Van Nostrand Reinhold Co. Ltd
Molly Millars Lane
Wokingham, Berks RG11 2PY, England

Van Nostrand Reinhold
480 Latrobe Street,
Melbourne, Victoria 3000, Australia

Macmillan of Canada
Division of Gage Publishing Ltd
164 Commander Boulevard,
Agincourt, Ontario M1S 3C7, Canada

15 14 13 12 11 10 9 8 7 6 5 4 3 2 1

Contents

Introduction

How often in the creation of a building does the safety and security of users receive adequate consideration? Generally, sufficient measures to alleviate building-related accidents and ill-health, crime and vandalism are not taken until an accident occurs, someone becomes unwell, security is breached or vandalism becomes rampant. Then the precautions are often deterimental to the appearance and utility of the building. Yet if safety and security are considered initially the elimination of hazards in an ergonomically sound design coupled with the hardening of criminals' targets and spatial design that will deter crime can provide unobtrusive safeguarding and at the same time preserve design from unwelcome modification by subsequent precautions.

This book shows how to provide for the safety and security of users by design and by material and component selection. Though it is desirable that provision is made at the design stage the contents are also a guide to the upgrading of existing work.

The common objective of safeguarding is recognised by covering safety and security aspects of the various parts and components of buildings in the same chapter. Nonetheless the division of chapters into sections ensures that readers whose main interest lies in one or the other areas will have no difficulty in following this throughout the book separately from the other area, though it is hoped that they will welcome the opportunity to examine the associated subject. Both safety and security are worthy of the same attention as that given to the other safeguarding function of building design, fire safety.

Ways that accidents happen and ways that vandalism and burglary are committed are illustrated because designers are more likely to take effective steps to improve safety and security if they know what might happen than if they are merely given a list of 'dos' and 'donts'. As the Chinese proverb says 'I see and I remember'.

Reference is made to British and American publications and standards that encompass good safety and security practice. The problems, and solutions, do not differ much in the Western World — and elsewhere for buildings designed in the 'international' style.

At the end of the book are checklists giving brief descriptions of

safety and security requirements in building design with reference
to sections of the book where full details are to be found.

May I hope to secure the cooperation of readers in reducing the
considerable distress and loss caused by accidents and crime.

Acknowledgements

Graham Locke, who represents the Royal Institute of British
Architects on the Child Accident Prevention Trust, read the draft
of the safety part of the book, and Chief Superintendent Brian
Ridd, Director of the Home Office Crime Prevention Centre, read
the draft of the security part; both made many helpful suggestions
that I was pleased to adopt. To them and to the many other people
who helped me in the research and writing I give my grateful
thanks. If there are any errors or important omissions the fault is
mine alone.

Extracts from British Standards are reproduced and adapted by
permission of the British Standards Institution, 2 Park Street,
London W1A 2BS, from whom complete copies of the documents
can be obtained.

Chapter 1

The Nature of the Problem

1.1 The situation today

A vaccine has removed smallpox — an old scourge — from the earth. Accidents — the modern epidemic — cannot be so simply extirpated. Accidents are no simple 'ill', they come upon us in a multiformity of ways and require remedies as numerous. One of the remedies is safe building design; it can eradicate some accidents completely — children will not fall from balconies if these are completely enclosed by railings that children cannot squeeze through. Other accidents, like falling down stairs, will always happen but their incidence can be reduced by considered design. Even the unfortunate case of smallpox which led directly to one death and indirectly to another in Birmingham in 1978 might not have occurred had the design of the building been different. It is thought most probable that the virus reached its victim through a service duct.

Security of buildings was once the concern only of the rich, the houses of ordinary folk were so flimsy that illegal entry could easily be obtained by breaking in through the walls. This is why we use 'house breaking' to describe the entering of a building illegally to commit a felony. Burglary in English law formerly meant house breaking at night, now it means house breaking at any time. It is a disturbing occurrence in any building, in a private house it can be most distressing, especially if accompanied by savage and obscene vandalism; even severe mental illness can result. It may be that the threat people feel is less than the real risk but today people do feel personally threatened, and they need the assurance of security in their own homes. In commercial premises robbery and theft have reached extraordinary proportions and building security must increase to combat the menace. Buildings are not well designed unless they give good protection against thieves and assailants. If today's accidents are an epidemic, crime is a growth industry.

1.2 The integration of safety and security

Often the requirements of safety and security in buildings are in conflict: ease of escape in case of fire and the need to keep out intruders is the example that always comes first to mind. There can also be conflict between safety of one kind and safety of another. If window openings are high enough above the floor to prevent children climbing up and falling out they may impede or even prevent escape in a fire. Also it is all too easy to remove one risk and create another: 'Every solution creates a fresh problem'.

Not long ago new houses were commonly given stair balustrades consisting of one or two longitudinal members with wide gaps between them. To remove the danger of children falling through the gaps a requirement was introduced into the building regulations of England and Wales that it must not be possible to pass a sphere of 100 mm (4 in.) diameter through the balustrade. This caused additional members to be put into balustrades, so forming a ladder with rungs 100 mm apart. Children found they could easily climb up this — so they did.

Climbing horizontal rails is a form of vandalism, though not vicious. Paintwork and perhaps the rails themselves will be damaged. If outdoors, the damage will accelerate deterioration of the paintwork. In this example the requirements of accident prevention and prevention of vandalism coincide. Unfortunately the issue is seldom as explicit. The designer has to decide what is best in the circumstances, but first he must attempt to see the many aspects of safety and security as an integrated whole.

1.3 The incidence of accidents

Accidents are seen as a modern epidemic only because other causes of death have declined. In Britain today, as in most other developed countries, one of the most likely ways of dying between the ages of one and thirty-five is from an accident. In this age group road accidents predominate. In the older age groups home accidents are more prominent; over 70% of accidental deaths in the home are to people over sixty-five. However children suffer too, one third of fatal accidents to children occur in the home, and accidents are the most common cause of death in children between one and fifteen years of age. The chance of a person dying from a home accident before reaching the age of sixty-five is believed to be between one third and one half that of dying from a road accident.

Fatalities are but the tip of the iceberg. Home accidents cause over two million attendances at hospital each year. One quarter of the accidents happen to children under five, and these involve stairs more than any other item. One- and two-year olds are particularly vulnerable. Fortunately children have outstanding powers of recovery and the death rate is low. In the over-sixty-fives the high toll of deaths occurs in a mere one-eighth share of all home accidents.

In Britain most of our knowledge about home accidents comes from the Home Accident Surveillance System (HASS). This was set up by the Government to ensure that consumer products did not produce undue hazards in normal use. Accident and emergency departments in twenty hospitals in England and Wales provide the information. They obtain it from patients coming to them for treatment following an accident in and about the home. Each year half of the hospitals are replaced by others so that sample bias is minimised in the long term. Among the data recorded, and subsequently put on a computer file, aspects of direct relevance to safety in building design are:

- A brief description of how the accident occurred.
- The products or features of the home involved.
- Where in the house or garden the accident occurred.

The Home Accident Surveillance System started in 1976 following a feasibility study in 1973—4. Before then, in 1972, a similar system had been established in the U.S.A. to provide national estimates of the number and severity of injuries associated with, but not necessarily caused by, consumer products and which were treated in hospital emergency departments. Known as the National Electronic Injury Surveillance System (NEISS) it was based on a statistical sample of 119 hospitals. It replaced two other systems that had been collecting injury data from hospital emergency departments. In 1978 NEISS was updated and improved, and the sample was revised to be statistically representative of the entire United States and its territories.

Building design is not, of course, involved in all home accidents, nor are all accidents where building design is involved confined to the home. Accidents similar to those in the home occur in other buildings, but their incidence may be different. In workplaces the problem of building hazards is eclipsed by that of industrial hazards. And in the familiar, built environment of the home, and perhaps workplace and school, people become accustomed to idiosyncrasies of design and learn to avoid hazards they are aware of. In strange buildings they tend to take more care but fall victim when their attention is distracted.

Building features have a frequent involvement in accidents in the home. Stairs, floors and doors make up about one fifth of the total number of items involved. Involvement does not mean that they are at fault or that they cause the accident. Whatever the cause of a fall the victim is likely to finish up on the floor, hence when asked what items were involved he is likely to mention the floor, especially when it was contact with it that injured him. However, research in Sweden showed that the main cause of at least one quarter of the accidents on stairways was the stairway itself.

Hard evidence is not available on the cause of most accidents. The supposition that design plays an important part receives support from the high incidence of accidents involving building features, reinforced by everyday experience. We have all stumbled over an unexpected step; we have heard how people walk into glazed panels not knowing they are there; the design for the effective guarding of balconies has already been mentioned. There are other more subtle dangers; after the introduction of new materials or new designs of components it is not uncommon for small epidemics of accidents to break out — people slipped on smooth vinyl floor coverings and children fell out of reversible pivot-hung windows.

1.4 Safety levels

How far should we go in providing safety in design? Nobody wants to live in a padded cell. I suggest we can look to structural safety for an assessment of the degree of risk requiring action. Buildings are constructed to withstand damage from wind up to the maximum gust likely to be exceeded on average only once in fifty years. This is felt to be a reasonable risk and on it the building regulations (for England and Wales, and Scotland) in respect of wind loading were based. This century the most extensive wind damage in Britain was done by the great gale of 2 January 1976. An estimated one and a half million houses were damaged. None of them blew down completely, even though, as the regulations had only been in force for ten years, most of them were probably built before legislation was felt necessary. The damage was mostly in the form of slates and tiles stripped from roofs; the dislodging of bricks and pots from chimneys, sometimes the demolition of stacks; partial destruction of inadequately tied-in gable walls; breakage of glass in windows. So it is mostly this sort of damage that wind might cause to houses once in fifty years.

Here perhaps is the yardstick by which to measure the acceptability of an item of building design in terms of person-safety. Over a period of fifty years is it likely to cause more than one accident requiring the victim to attend hospital? If it is, then it must be dispensed with or made safer. In the absence of detailed records like those used by the Meteorological Office to predict wind gusts the designer must make some sort of subjective judgement. The scale shown in Fig. 1.1 might assist in this; for example, if the risk of an accident is 3 or under then the design is definitely unacceptable. An assessment of 4 might be 'An accident could conceivably happen'. This should be regarded as bringing the risk into the 'more than once in fifty years' category.

Some items, like stairs, are intrinsically dangerous, so the assessment must be at least 3 on the scale. Here the designer must

There'll be an accident here before long		I can see the possibility of an accident here		Never in a thousand years
1	2	3	4	5

Figure 1.1 Assessment of the risk of accident.

aim to make the item as safe as possible. The yardstick suggested is for a minimum level of safety. Progressive levels are found in some safety recommendations. For instance, the tests of BS 6206 and ANSI Z 97.7 put safety glazing into three classes (see section 6.3): a classification that is supported by guidance in BS 6262 on the class of glazing to use in different locations. This guidance is based on the number and likely behaviour pattern of people using a building. Another example is found in the approved documents of the building regulations where there are different requirements for stairways in different classes of building.

It has not been found practicable in this book to make recommendations for safety at different levels, in other words, to have grades of safety. However it has been necessary to distinguish safety requirements that are specific for children or old people. It can generally be assumed that where design ensures the safety of children and old people it will ensure the safety of other age groups, but there are buildings where the requirements for children and old people need not be catered for, either because children are not allowed into the building or because old people are not likely to venture there — a factory for example. For this reason, in the safety checklist at the back of this book requirements that are specific for children are marked with a C and those that are specific for the elderly are marked with an E. In most buildings the requirements of children and the elderly should be catered for, certainly they should be incorporated in domestic buildings. Even in old people's homes design must provide for the safety of children, so that they can visit their grandparents in safety.

1.5 The cost of safety

Improving safety in building design need not necessarily cost more. Often it is merely a matter of repositioning features in relation to one another. Sometimes the use of a safer and more expensive material will be called for, e.g. a different floor covering. Sometimes the construction will need to be more extensive, e.g. a higher balustrade. Sometimes additional items will be required, e.g. another handrail. These are small items not to be compared with the millions spent on fire safety or as protection against wind damage. And when viewed in the light of what expenditure on such measures may save, not only in money but in terms of human misery, person-safety will be seen as being extremely cost-effective.

Consider the plight of a child injured and taken to hospital; there, in a strange and frightening place, to be handled by strangers who hurt him; probably to be kept there, away from home for the first time; later coming home, handicapped or scarred for life; becoming the object of special attention and so incurring the resentment of brothers and sisters; at school an oddity having to endure the taunts of thoughtlessly cruel contemporaries. One parent blames the other for the accident and ultimately, under the burden of guilt compounded by the strain of caring for the child, the marriage breaks down. Who knows what the full cost may be?

1.6 Responsibility of the designer for safety

When a number of people are killed and injured in a single incident such as a fire or the collapse of a building there is a public outcry, official enquiries, perhaps new regulations. When people are killed and injured in far greater numbers separately, though the accidents would not have happened had the design of the building been different, there is no outcry, no enquiry (except a coroner's court). The solitary victim (or his parents) should have been more careful, he has contributed to his own misfortune, is the generally held view, whereas the mass accident was someone else's fault — the designer's perhaps.

In truth, a range of interacting variables lie at the root of all accidents and to improve safety all aspects of accident prevention must be pursued. As we shall see, where there is scope for mechanical improvement this should be one of the first measures taken. Unfortunately, designers often see mechanical safety as interfering with their concept of the feature or item. Le Corbusier exhibited this view when he designed the staircase shown here (fig. 1.2). Handrails would have spoilt the sculptured form which he allowed to take priority over safety considerations. When accidents happen in such cases, improvised remedies destroy the vision conjured up by the designer. His intentions would have been better realised by a design that was mechanically safe in the first place.

Figure 1.2 Style before safety: stairway in apartment Le Corbusier built for himself in Paris.

The designer must not allow the pursuit of artistic achievement or other objectives to obscure the need for safety. A mother expects her home to þe a safe place for a small child to play in. The elderly expect to be protected against their infirmities if they remain house-bound. The public in general expect not to be put at risk unnecessarily in the buildings they use. It is the designer's duty to meet these expectations. Le Corbusier's stairway was not for himself alone — he had a wider responsibility, other people would use it. Design that does not accommodate safety is not responsible design.

1.7 The incidence of security failures

Burglary accounted for one quarter of all serious offences recorded by the police in England and Wales in 1983. The total was 808 000 and over half the offences (432 000) were burglary in a dwelling. Many of the less serious burglaries are not reported to the police and not all that are reported are recorded. The British Crime Survey (BCS) (Home Office Research Study (1983) No. 76. London: H.M.S.O.) estimate of the number of household burglaries in 1981, the year the survey was carried out, was 726 000 compared with the recorded figure of 350 000. However many of the burglaries involved no loss, others were merely attempts (perhaps frustrated by good security design). The BCS estimate of households burgled with loss is 400 000 (± 70 000). In the U.S.A. police recorded rates are about 50% higher than in England and Wales, with possibly a smaller proportion being reported.

The 443 000 cases of criminal damage recorded in England and Wales in 1983 are only the tip of the iceberg of vandalism. The BCS estimate of the number of incidents of vandalism to household and personal property in England and Wales in 1981 is 2 650 000 or about 1500 per 10 000 households. Schools, shops and some other buildings face a greater risk. The data are not available to enable the cost of wilful damage to buildings to be calculated but it undoubtedly runs into many millions.

Burglary is only part of the problem of theft from buildings. Internal theft involving employees occurs on a large scale and building design can help to reduce it. Employees, disgruntled ones in particular, are also responsible for vandalism at their workplace. Intruders who are not bent on any premeditated wrongdoing but who are nonetheless likely to cause a breach of the peace can be a problem in educational buildings, especially when the buildings, with many entrances and exits, are only partly in use in the evenings. Their corridors are warmer than the streets for idle gangs to roam and loiter in.

Running counter to keeping people out is the need to keep them in, not in special buildings like prisons and psychiatric hospitals but in large factories where an employee can clock in and then sneak out for a day off without being missed.

1.8 The need for secure buildings

The odds given by the British Crime Survey for a 'typical' household being burgled are one in forty a year, which is less favourable than the one in fifty risk accepted for wind damage (section 1.4). However the risk of burglary varies with the locality; in the inner city the risk of a house being burgled is one in thirteen a year. Scottish rates for household burglary are broadly similar to those of England and Wales, while U.S.A. rates are about twice as high.

In view of the apprehension felt by most people about violation of their home, and indications that the rate of burglaries is increasing, there is a strong case for giving special consideration to security in the design of residential buildings. The case is still stronger where, because of the location or contents, a home is at greater than average risk.

Such statistical assessment is not possible for commercial and other non-residential buildings, however the rising crime figures make the case for security in their design.

An accident prevented is one less for the statistics, but unless the determined burglar is after something special if he finds one building too much of an obstacle he will move on to something more accommodating. However there is an element of opportunism in burglary, the criminal is tempted by easy pickings. During the exceptionally hard winter in Britain of 1981—2 there was a considerable decline in burglaries. Therefore it can be argued that a general raising of security will reduce the incidence of burglaries. The professional will raise his expertise but the majority of burglars are youthful amateurs lacking in technique and they will be more easily discouraged by good security design.

Much has been said about the underlying causes of vandalism, and of course if vandalism is a symptom the causes must be tackled. Vandalism is, however, definitely an opportunistic type of offence so removing the opportunity will reduce the incidence. The aesthetic worth of an architectural feature disfigured by vandalism will be much devalued so it is foolish to invite spoiling or misuse.

1.9 The cost of security

Improved security in the ordinary run of buildings does not necessarily add to the cost of the building. The avoidance of undesirable features may actually reduce costs. The trellis shown in the illustration (fig. 1.3) is an example; if it is going to be vandalised it will not improve the appearance of the building, which presumably is its function, so it is best left off. Sometimes, as with safety, security can be improved simply by adjusting the relative position of items. Above all, the full costs-in-use must be considered: replacement costs of articles stolen, repair costs, time spent by employees and police on prevention and apprehension of culprits, the psychological effect on the victims.

Figure 1.3 This trellis is the sort of feature that invites climbing and swinging from and makes vandals out of playful children. Deliberate vandals and burglars may use it as a way on to the roof. (Reproduced with permission from *The Architectural Press*.)

1.10 The designer's responsibility for security

In the case of buildings that will contain items of high value or that will be regarded as targets by terrorists the designer will be aware of his responsibility for security. For other buildings he has a responsibility not only to see that they are not made unnecessarily vulnerable to illegal entry and vandalism but to take active steps to provide good security.

Chapter 2

Fundamentals of Safety Design

2.1 Definitions

Safety of the person may be defined as a state of freedom from danger to health from accidents and disease. In normal parlance safety means a reasonable degree of that state, it being recognised that absolute freedom from accidents and disease is impossible in everyday life. Our special view of person-safety is confined to design that does what is reasonably possible to prevent accidents arising from the building environment (excluding fire) and to prevent diseases arising from building materials and services. For the moment our concern will be with the prevention of accidents; the prevention of disease, which is a slightly different problem, is left until chapter 5.

A minor difficulty in delineating an area of safety is that there is no generally accepted scientific definition of an accident. In general use, injury or damage is usually a prerequisite of an accident, though the word can be used to describe an unforeseen event with a favourable outcome — Columbus discovered the New World by accident. In accident research some definitions demand that an accident shall result in an injury, others accept damage as an alternative to injury, still others allow almost any unplanned event to be included. However, in the majority of research papers the words accident and injury are almost synonymous. They cannot be entirely so because deliberately inflicted injury is not regarded as an accident, also some accidents do not cause injury. For our purposes a strict scientific definition is not necessary, the common meaning, that it is an unlucky event causing injury, suffices.

2.2 Accident proneness and accident liability

But are accidents just bad luck? The concept of accident proneness postulates that some people have a persisting and stable personality characteristic that makes them more than usually likely to have accidents, perhaps because this satisfies some unconscious need to suffer. The concept finds ready acceptance by the public and the term accident proneness is part of everyday speech. People feel that they know persons possessing the characteristic, they see it exhibited by the accidents these persons have and associate it with some quirk of personality. There is danger in acceptance of the concept, it leads to the defeatist attitude that it is accident prone people who have accidents and that they cannot be protected by improvements in the environment. This attitude is probably founded on a fallacy, attempts to define scientifically an accident proneness characteristic of personality have not been successful.

Some of the difficulty lies in the nature of the data that have to be used when we try to establish some general case about this aspect of individual behaviour. A fatal accident cannot be repeated, a severe injury may necessitate a long period of immobility, a minor injury will make some people more careful, others more nervous. The chief difficulty is the small number of accidents contributed by each individual. However it is not disputed that some classes of people are more likely to suffer accidents and that some individuals are at times more at risk than normally. These classes and individuals have an increased accident liability, which differs from accident proneness in that it is not a persisting and stable state.

Children and old people are obvious examples of classes that have an increased liability to accidents, in their case because of physical and psychological factors associated with youthful or advanced age. With individuals, all ages may find themselves subjected to strain, leading to a disruption of rational thinking and decision making — they do 'silly' things or become 'careless'. The turmoil and confusion pertaining in some home environments, work problems, marital problems, ill health, moving house, uncomfortable working conditions and a variety of other stressful circumstances are associated with an increased liabilty to accidents. The period of strain may be prolonged, as with long-term unemployment, or it might be of short duration but still long enough to play a part in an accident — as with a candidate for a post who left a gruelling interview, walked down the stairs and straight through a floor-to-ceiling window.

Alcohol is also a cause of accident liability. A study of 280 deaths from home accidents in the 15—64 age group showed that alcohol was a major factor of twenty-nine of the sixty-eight fatal falls in the sample. (Anon (1980) 'Personal factors in domestic accidents: prevention through product and environmental design'. London: Consumer Safety Unit, Department of Trade.) Thus intoxication as well as all sorts of strain, unruliness in children, bad temper in adults, infirmity, distractions and tiredness may lead to accidents in situations where the building designer did not envisage an accident risk.

2.3 The accident chain

Social and economic factors also play a part in accidents, they influence the kind of buildings erected, their design and construction. These factors, as well as personal factors, constitute a series of influences and events leading up to an accident. Such linked factors may be construed as an accident chain. Where the chain starts is beyond definition. 'Does the flap of a butterfly's wings in Brazil set off a tornado in Texas' was a meteorologist's way of illustrating the point that in the atmosphere the smallest of events will in principle affect everything else. So it is with accidents and indeed, one supposes, with the whole of life. To be practical, consideration of the causes of accidents has to be confined to those causes proximate to the final event.

When accidents are analysed it is convenient to form one chain of events involving the victim directly and other chains of related events which come together to bring about the accident situation. Such an analysis of a fatal accident is shown in fig. 2.1. The chains show that an accident may involve many factors of building design. The decision to build high, the use of reversible windows to allow cleaning and maintenance from inside, the design of the safety stay and the provision of a wide internal sill or ledge that a child could climb on to all contributed to the accident. The proximate factors were the wide sill and the design of the safety stay. A mechanical contrivance like the stay is always likely to fail to operate and reliance must not be placed on users checking it, probably it will never occur to them that they should. Design needs to be inherently safe. (How this might have been achieved with this window is considered in section 2.7.)

It is pointless to blame the mother for leaving the child alone, she expected the child to be safe within the flat. Building design should prevent tragedies like this by providing a safe environment. Personal, social and economic factors make an important contribution to accidents but they are not so easily dealt with, and measures taken are seldom as effective as those dealing with the physical setting or agent of harm. In the hot, oppressive New York summers, around two hundred of the city's children were killed or injured every year in falls from open windows. The provision of window guards coupled with a safety campaign resulted in a 35% decrease in fatal falls over three years, and *no falls were reported from windows where guards had been installed*. (Spiegel, C.N. and Lindaman, F.C. (1977) 'Children can't fly: a program to prevent childhood morbidity and mortality from window falls.' *American Journal of Public Health*, **67**, 1143—1147.)

Other instances outside building design where tackling the physical setting or harm agent has been shown to be the most effective method of accident prevention include:

The reduction in fires caused by paraffin heaters when these had to be made self-extinguishing on tipping over.
The reduction in deaths, particularly among children, following the issuing of regulations on the resistance to flame spread of nightdresses.

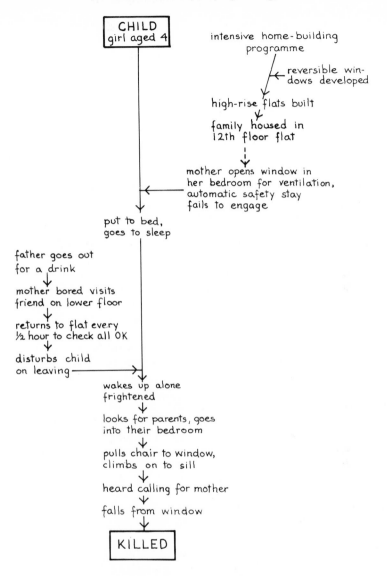

CHILD
girl aged 4

intensive home-building
programme

← reversible win-
dows developed

high-rise flats built

family housed in
12th floor flat

mother opens window in
her bedroom for ventilation,
automatic safety stay
fails to engage

put to bed,
goes to sleep

father goes out
for a drink

mother bored visits
friend on lower floor

returns to flat every
½ hour to check all OK

disturbs child
on leaving

wakes up alone
frightened

looks for parents, goes
into their bedroom

pulls chair to window,
climbs on to sill

heard calling for mother

falls from window

KILLED

Figure 2.1 The chain of events that led to a fatal fall.

The striking reduction in child poisoning brought about by regulations requiring that all aspirin and paracetamol dispensed for children should be in a child resistant pack.

2.4 The perception of hazards

We know little about how people perceive hazards in buildings. Children have to learn what is dangerous. Provided their injuries

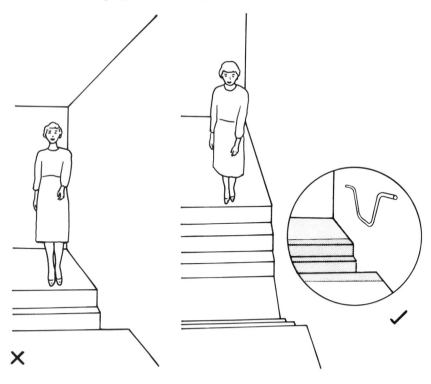

Figure 2.2 A child's eyeview of an approaching stairway hazard compared with that of an adult — and the cues needed.

are no worse than superficial wounds like scratches, abrasions, split lips, bruises and blistered fingers they are accepted as part of the learning process. What we do not know is whether the lessons have been properly learnt. Adults often fail to anticipate dangers, as in the case of the safety stay previously mentioned. However if we keep the safety of children to the forefront we shall generally provide for the more inattentive adult.

Anything that can be climbed, squeezed through or swung on will be so treated by children, the designer must therefore be on the alert for such items. Small children are not much aware of heights, a two-year old boy dropped his teddy bear to the ground from a balcony 3 m (10 ft) up then climbed over the railings (which should not have been possible) to retrieve it.

Things look different when you walk nearer the ground. Children do not see dangers from as far off as adults. A child's view of a stairway containing two short flights with a landing between them, compared with an adult's eye view is shown in fig. 2.2. An accident to a five-year old boy occurred on a stairway like this. Both floor and stairway were covered with the same carpet so the stairway did not stand out clearly. The boy was walking with an adult and looking straight ahead, he did not see the stairway at all and fell down it. The adult had more time to perceive the gap in the floor and become prepared for the descent.

Warnings and cues are necessary to alert building users to dangers, especially where there is a change of level. A hospital had to put gates across the head of a stairway like that shown in fig. 2.3: it was often mistaken for a corridor. People were liable to find out they were wrong only when they fell down the stairs. Ways of stimulating attention before stairways are reached are considered in section 11.2. Cues are needed and distractions near danger points must be avoided. Mirrors are particularly distracting, people look at themselves instead of where they are going, and additionally mirrors can give confusing and false information (see fig. 2.4).

2.5 The ergonomic approach to accident prevention

For building design to provide a well thought out safe environment the application of ergonomic principles is called for. The term ergonomics (from the Greek *ergon* (work) and *nomos* (law)) is accepted in Britain to describe the study of man in relation to the machines he uses. From initial studies of men using mechanical contraptions at work the science has grown to embrace man's relationship with machines in the broad sense of any contrivance of mutually adapted parts working together. It therefore includes buildings — both the components and the environment created. In North America the range of ergonomics is covered by the terms human engineering, human factors science and engineering psychology.

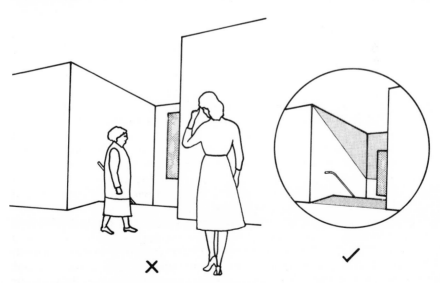

Figure 2.3 In the absence of cues to the presence of a stairway the user is not prepared for the danger ahead; warning signs, as shown in the insert, will be subconsciously interpreted.

Figure 2.4 Distractions and confusing information at a danger point (reproduced with permission from Morley von Sternberg).

The aim of ergonomics is to tune machines to efficient human operation. In industrial and military applications, high productivity and speed of response are important aims. In almost all applications comfort and lack of fatigue are important, in the home they are paramount. Accidents are incompatible with efficiency and comfort, ergonomic design is therefore safe design. The ergonomic view of an accident is that the situation is at fault, not the victim.

The adoption of this attitude and the application of the ergonomic practice of designing from the man out, observing the requirements and limitations dictated by man's body size and physical performance (described in chapter 4) and the avoidance of the hazards depicted in later chapters will provide the framework for safe design. Its use will prevent dangers arising with new, at present unconceived, designs and avoid the unwitting creation of fresh hazards when eliminating old hazards.

In all situations the requirements of safe design are seen more clearly if they are divided into primary safety and secondary safety. Primary safety aims at freedom from accidents, secondary safety at ameliorating the immediate effects of the accident. With a floor surface in, say, a nursery school, a non-slip surface would give primary safety, rounded corners on columns in situations where falls might occur would give secondary safety. There is another level of safety but it is outside our concern — tertiary safety seeks to ameliorate the long-term effects of accidents.

2.6 Safety appraisal

Accident data are collected with the object of identifying common factors and patterns and of taking or recommending some general corrective action. However accidents occur in such multitudinous ways that it is difficult to make order out of the data. Various classifications of the way accidents have happened, the part of the body injured, etc., are made, mostly in occupational safety and some of them of doubtful value. However two of the classifications, 'type of accident' and 'hazardous environmental conditions', have sub-divisions with headings that are of use to the building designer. Examples are accident types categorised as fall from elevation, fall on same level, struck by, struck against, caught in, caught under, caught between, and so on. Environmental conditions have categories that include excessive noise, inadequate clearance, inadequate ventilation and improper lighting. The use of these headings when appraising a design for safety focuses attention on possible hazards.

The checklist (fig. 2.5) for use in design appraisal covers types of accidents and environmental conditions. Also included are the bodily characteristics we shall consider in chapter 4, and items to remind us that the clothes and jewellery people wear can increase the possibility of accidents. Just as people are adorned with clothes and jewellery, so buildings are adorned with drapery, carpets, pictures, light fittings, etc. Floors are polished and they get wet. All forms of maintenance, cleaning, re-decoration and repair must be considered in the appraisal, and thought must be given to safety during construction of the building — can a large sheet of glass be installed safely, for example. If inherent design for safety is not possible and recourse to safety devices is the only solution, question how the devices will function over the lifetime of the building.

2.7 A case study

'Histories make men wise' and there is much we can learn with hindsight about the need for safety appraisals from the history of window design for high-rise flats. So that flat occupiers could clean their own windows, reversible horizontally pivoted windows were extensively used in the high-rise building programme of the 1960s. These made the outside of the window accessible from the inside but inevitably at the same time made the dangerous outside space accessible. Attempts were made to provide for safety, with

Both the item and its vicinity are to be appraised (i) in use, (ii) under maintenance. Account is to be taken of anything used with the item, e.g. curtains with windows, polish with floors.

Examine the design and question the possibility of any of the following kinds of accident occurring
(a) Could persons fall from elevation, or be caused to fall on the same level?
(b) Could persons be struck by the item or part of it? Is there anything they might strike against?
(c) Is there anything that persons or their clothing could be caught in, under or between?
(d) Could persons come into contact with electric current, temperature extremes, radiations, caustics, toxic or noxious substances?
(e) Is there anything that could cause injury or rubbing or abrading?
(f) Could use of the item cause a person to overexert himself?
(g) Could physical impairment of breathing or other bodily function be caused?

When examining the design for possible accidents consider the following with special attention to children and old persons
A. *Characteristics of item* Is the possibility of an accident increased if the item causes persons to stretch, stoop or bend, or if the item is sat on, stood on, walked on, run on, climbed up, squeezed through, held, moved or lifted?
B. *Bodily characteristics* Does the design make proper provision for, and avoid excessive demands on persons in respect of their height, build, reach, stride, gait, balance, strength and grip?
C. *Dress* Is the risk increased for persons wearing long sleeves, other loose clothing, necklaces, bracelets or ill-fitting footwear, or if persons go barefoot?
D. *Environmental conditions* Is the visibility adequate in daylight and in artificial light? Is the artificial light switch accessible in safety in the dark? Is ventilation adequate? Will ambient temperature have any effect when it is hot or cold?

When safety cannot be inherent in design and safety devices have to be used, ask these questions
(i) What will be the outcome of failure to use the device as intended?
(ii) How could the device malfunction?
(iii) What is the life expectancy of the components?
(iv) What will be the effect if maintenance is neglected?

Figure 2.5 Checklist for safety appraisal of design.

unanticipated consequences that serve to illustrate the need for a disciplined approach to reveal the whole of the safety problem.

Millions of reversible horizontally pivoted windows (fig. 2.6) were built into the flats. Over twenty British manufacturers produced four to five million steel windows complying with BS 990 (*Steel Windows Generally for Domestic and Similar Buildings*). They were cheap and many were exported, along with their problems. The windows were intended to be opened slightly for ventilation and turn over completely for cleaning the outside glazing surface from inside. A British Standard requirement was that the initial opening for ventilation purposes be limited by means of a safety stay. Originally the limitation was 15°, this gave a gap of 140 mm (5½ in.) at the sill for a window 1050 mm (3 ft 6 in.) high; with 'give' in the window and in the stay the gap would increase by about 12 mm (½ in.) under pressure. Several children up to five years of age contrived to get through these nominal 140 mm gaps, and the larger gaps of taller windows, and fell to their deaths. The recommended maximum ventilation opening then became 100 mm (4 in.) for all windows — where it remains today.

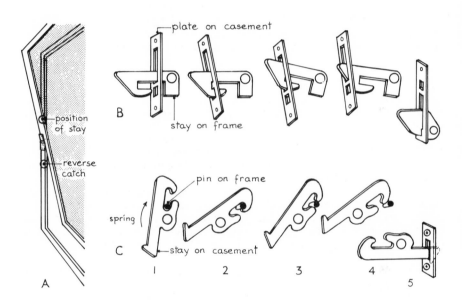

Figure 2.6 (A) Reversible horizontally pivoted window; (B) drop down safety stay; (C) spring-loaded safety stay. *Operation of stays:* (1) Window closed; (2) window open for ventilation; (3) released for reversing window; (4) window closing; (5) window reversed. *Possible failure of stays:* Stay fails to drop as at B.1 when window is closed but remains up as at B.3. Stay sticks in release position as at C.3 when window is open for ventilation, thus there is no check on the initial opening.

At first a manually operated stay was used to limit the initial opening of the window. It had to be deliberately applied when the window was opened for ventilation and released when it was reversed for cleaning. If the stay was not engaged while the window was open for ventilation the windows would open fully in response to a slight push. In an attempt to compensate for human failings, manufacturers developed safety stays that automatically stop the window going beyond the 'safe' initial opening, they have to be released when the window is reversed but on closing they re-engage automatically. There are various designs, two of which are shown in fig. 2.6. Either the stays drop into place or they are held down by a spring. They can fail to engage if they foul the weather stripping or are bent or painted. Children have climbed up on to window ledges, in play or to look out of the window, the window has been open for ventilation without the stay engaged, the children have pressed against the window and it has given way.

Lockable safety stays have been developed, the window cannot be reversed without unlocking the stay which is re-locked on closure. However, after a few changes in tenancy the keys get lost or the new tenants do not know what the keys are for. In one 'foolproof' lock the key cannot be removed until the initial opening device is engaged; the device could fail to operate if the lock spring broke, so two springs are provided. What will happen in the course of time remains to be seen. The basic problem is that there is no mechanical way in which these safety devices can be made to allow release for reversing and also to fail safe.

Before a small epidemic of accidents drew attention to the dangers, many reversible windows of wood construction were made without any restriction on the initial opening whatever. Also wood and metal windows containing a side-hung light adjacent to a pivoted light were used with a safety stay on the pivoted light but with nothing on the side-light to limit the distance it opened.

As a large reversible window is opened fully a yawning gap opens up in front of the user, causing vertigo in some people. When turned over completely the window must be fastened in the reversed position to hold it secure while being cleaned. The catch that secures the reversed window causes as many problems as the safety stay and is just as likely to malfunction. When all works properly up to this stage of the cleaning operation there is still the possibility of an accident if the person cleaning the window has to stand far off the floor to do the job, and cleaning side lights safely from an open pivoted window is virtually impossible (fig. 2.7).

A safety appraisal, using the checklist, of the windows of this history reveals:

- People can fall from elevation both inside and outside (a).
- The possibility of an accident is increased by them having to stretch, and by them (children particularly) standing on the window ledge (A).
- It makes excessive demands in respect of height (toddlers are not tall enough to see out) (B).
- Balance may be affected by vertigo (B).
- Over-riding of the safety stay for increased ventilation raises the risk and consequences of an accident considerably (i).
- The safety stay can malfunction if bent or binding on something (ii).
- Fifteen years is a reasonable estimate of the working life expectancy of a safety stay (iii).
- Neglect of maintenance will result in the safety stay becoming

Figure 2.7 The virtual impossibility of cleaning a side light in safety when obliged to reach over a pivot hung light.

Figure 2.8 Inherent safety. When the need to stand off the floor is removed a window is safer for children and adults without reliance on safety devices.

inoperative in a short time, possibly after the first re-painting of the window (iv).

Obviously reliance must not be placed on a safety device with these windows, the design must be as inherently safe as is possible (fig. 2.8). With this the bottom of the opening light is high off the floor but its top is not at a height that requires much aid to reach; fixed side lights are not used; there is no window ledge; a fixed laylight of laminated glass brings the sill low enough for toddlers to look out without climbing on furniture. Vertigo, it is hoped, is prevented by the laylight. A safety stay is still necessary but its failure will be of less consequence.

Chapter 3

Fundamentals of Security Design

3.1 Aims

In making provision for security in building design the aims are:

- To prevent unauthorised entry.
- To impede the removal of stolen goods.
- To reduce vandalism directed against the building.

Foreceful entry into most buildings requires no very special skill; almost any person turning to crime can attempt burglary, it is an anonymous crime usually requiring little determination or perseverance and the risk of detection is low. However, careful building design can ensure that the criminal is less likely to attain his objective. No-force unauthorised entry into commercial and industrial buildings is simple unless restricted by some form of access control. Pilfering by employees can be limited by means which prevent them from entering certain parts of the building, and by planning that prevents them from leaving the premises unobserved. Vandalism can be reduced by robust construction and other measures.

3.2 Grades of protection

The most common form of crime to guard against is burglary. Table 3.1 gives a rough classification of burglars and the sort of building they attack. The opportunist and minor criminal is likely to be young and inexperienced, his low-skill approach is directed against any easy target no matter how small the reward — garages and outhouses are not beneath his attention. He likes to attack buildings he knows, hence schools, colleges, youth clubs, benefit offices and dwellings in his own neighbourhood, are attacked because they are familar buildings, though they may contain little of value to a thief.

Professional and semi-professional criminals engaged in what is described in table 3.1 as deliberate and organised crime may operate

Table 3.1 Burglary and buildings

Class of crime	Type of criminal	Kind of building attacked	Comments	Grade of security required
Minor and opportunist	Local delinquents Social inadequates	Ordinary domestic property Small shops and offices	Seeking cash, cigarettes, wines and spirits, home entertainment items, anything that can be turned directly into cash ('It fell off the back of a lorry') Easily deterred	Basic
Deliberate	Professional and semi-professional	Opulent houses Hotel rooms Warehouses Department stores	Plans attack, with varying consideration of risks and probable rewards Fairly determined	Strong
Organised	Professional	Banks Cash offices Furriers Jewellers	Plans carefully, arranges cover, secures outlets Will go to extreme lengths to achieve objective	Maximum

at the 'craft' level using well practised techniques in a routine way for crimes such as housebreaking and thieving from hotel bedrooms or at a higher level they may carry out 'project' thieving. This is the type of operation where a gang is assembled to attack a specific target. Experts on alarm systems, cutting techniques, etc., are brought in and the assault and subsequent operations are carefully planned.

There are, of course, other ways of classifying crime and criminals, for instance Maguire (Maguire E.M.W. (1982) *Burglary in a Dwelling*. London: Heinemann Educational Books) determined three patterns of behaviour of persistent burglars based on value of items taken — low level, middle grade and high level. He suggested that these corresponded to amateur league, second division and first division, designations that are also appropriate to the criminals considered in our grades of basic, strong and maximum.

Stratifying security into three grades enables items and design features to be graded according to the protection they give. Of course the grading is merely a guide, no great precision can be achieved. Strong security, for instance, will encompass graduations varying from fairly strong to very strong. However even a rough grading of items and features assists the designer to attain the desirable objective of the same level of security throughout the zone of the building that is to be protected. To leave a weak point in the general level of security or to make overprovision in some places is uneconomic, but this is not to say that there will not be different security zones in the same building; offices containing electronic typewriters and other articles likely to attract burglars may warrant a higher grade of protection than public areas of the building, though generally it is preferable to raise the grading of low-risk areas for greater protection of adjacent higher-risk areas.

A single feature may warrant a higher grade of security than the rest of the building if it is in a position where a criminal can work unobserved. For example, if a window must be placed in a passageway or other secluded location then it may be necessary to protect it with bars or a grille; the back door of a house with a walled garden needs to be more secure than the front door on a busy, well lit street.

It must be recognised that as criminals become more skilled so security has to be raised, and when defensive technology gets the upper hand criminals change their tactics. They look to softer targets, bandit resistant screens in banks turn them towards building society offices and sub post offices, similarly target hardening of commercial premises turns them towards household burglaries. Improved security in building design may also transfer the risk of attack from buildings to people: kidnapping the manager and forcing him to open the premises, for example.

If a robbery in which the criminal confronts his victim is likely, the need is for maximum security in physical terms coupled with control of access to target areas. Violence without robbery, of the unpremeditated kind, is likely whenever people find themselves in desperate circumstances. Officials who allocate benefits are sometimes attacked by exasperated claimants. The officials then demand screens for their protection. Compliance with this demand leads to a need for access control to ensure that the claimants remain on the right side of the screen. Premeditated violence in the form of terrorist attack must be considered likely in buildings used for government purposes or occupied by persons in the public eye. Attacks are also made against buildings where people congregate — restaurants and air terminals have suffered in the past. A terrorist's technique differs in some ways from that of the usual criminal, he may be prepared to die for his cause (the 'kamikaze' attack) and he may not need to enter a building in order to attack it. Maximum security with special precautions is called for where there is a probability of terrorist action.

In certain buildings the population must be considered hostile. Prisons and detention centres are obvious examples, but schools and colleges, barracks and some workplaces will hold large numbers of users who feel no responsibility for what happens to or in the building. If not overtly hostile they are at least indifferent. Two other classes of users might be distinguished — responsible and security conscious. Responsible users will report wrongdoing, security conscious users will challenge strangers and take active steps to ensure that security is maintained.

The effects of gross environmental and social influences on burglary are shown by the way that the residential burglary rate increases in the progression: small towns/medium-sized towns/large towns/large cities, with perhaps eight or ten times the risk in a large city compared with country districts. An exception is a small town within an easy car journey of a large city, where the rate may be relatively high. Most elements of the built environment, such as streets, parks, open spaces, are factors in crime. Heavy vehicular traffic through a residential area can lead residents to feel that they have no control over the neighbourhood and thus affects their sense of responsibility. Open access to large blocks of flats has a similar effect.

3.3 Vandalism

Many anti-vandal measures coincide with anti-burglar measures, but the destructive and apparently objectiveless nature of vandalism necessitates special consideration. As an aspect of behaviour, vandalism is not easily distinguished. Young children's play (e.g. dropping stones down a gully trap), older children's games of skill (e.g. who can break the most windows) and such pastimes, coupled with hard wear and tear by uncaring adults, shade into wanton defacement and destruction which in turn shade into destruction in the pursuit of gain (instrumental vandalism) such as stripping copper pipes. The play framework makes it largely a crime of the young. It tapers off in the teens when other interests develop — an interest in the opposite sex, perhaps. Unfortunately this tapering off is counterbalanced to some extent by the vandal's increased strength and reach.

As with an accident, an act of vandalism in a building, whether it be rough play or wanton destruction, results from a chain of germane events. A broken home, lack of educational opportunities, unemployment, inadequate facilities for leisure and recreation, and other factors, are said to be linked to the act. Building design may figure in more than one part of the chain, as it did in the accident analysed in section 2.3. In this, the high-rise building programme and the questionable practice of housing families in flats were important influences, but at the end of the chain lay a design detail that was seen to have made the accident possible. So it is with vandal resistant design, whatever the underlying social factors and planning decisions, at the point of contact is an item with unequivocal involvement. General and large scale design features must take account of the nature of the vandal and aim to influence his behaviour but the assailed feature must stand up to his attack.

Generally there are three proximate prerequisites to an act of vandalism committed in or around a building:

- The area is not under surveillance, official or unofficial.
- There is no apparent owner.
- There is already some damage in the area.

The first of these is an obvious prerequisite, the vandal does not want to be caught. The second is probably most influential with children. The third is important because it would appear that when the end of the chain is reached a trigger or release is required to spark off the destructive impulse. The appropriate grade of security to adopt as defence against vandalism may be determined as indicated in fig. 3.1. In addition, some special measures directed against vandalism alone will be necessary — graffiti deterrence for example.

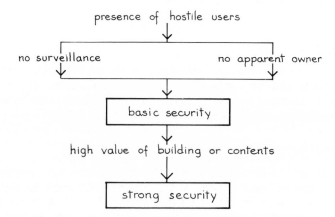

Figure 3.1 Determining necessary grades of security for resistance to vandalism.

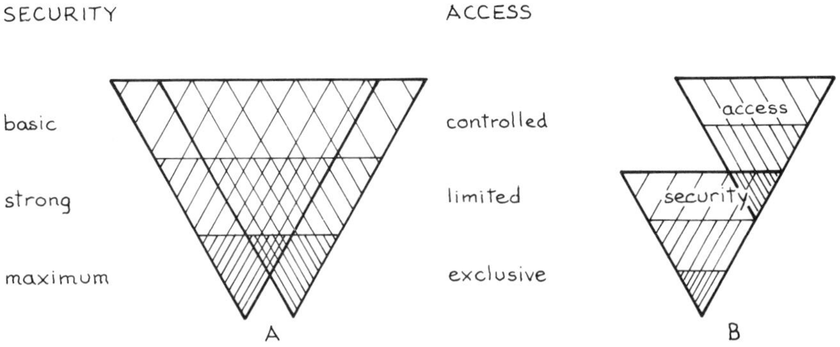

SECURITY   ACCESS

Figure 3.2 (A) Grades of security matched by levels of access control; (B) access control may be at a higher level than security, but the reverse is unlikely to be required, or be practicable.

3.4 Access control

The object of access control is to restrict entry to specific zones to selected persons: to let in only those persons that you wish to let in. It is frequently not necessary to have a high grade of security in so far as strength is concerned, access control is a form of self-policing carried out during working hours and unauthorised persons attempting entry will not draw attention to themselves by the use of force. As with the grades of security we have used, our reference framework for access control also has three divisions, these are:

- Controlled access.
- Limited access.
- Exclusive access.

At the lowest level of security, entry is controlled to exclude the general public. At the next level, entry is limited to selected, cleared personnel. At the highest level, all but the most essential and trusted personnel are excluded. For controlled access a turnstile may be sufficient, for the higher levels of control, locked doors will be necessary. The doors may be opened by conventional keys, push button digital controls, card access readers or combinations of these and similar lock operating devices (see chapter 13).

The relationship between levels of access control and grades of security is shown in fig. 3.2.

3.5 General defensive measures

Based on research in the City of New York, in 1972, in the book *Defensible Space*, Oscar Newman put forward his ideas of the value in crime prevention of a sense of territoriality among residents of public housing and opportunities for surveillance by them. He also expounded on the corrupting influences of design that gave public housing an institutional image, and on the benefit obtained from the juxtaposition of residential areas with 'safe' zones.

The book gained wide acceptance of the concept of defensible space but not all Newman's conclusions are now accepted and some are not applicable outside the U.S.A., where there are substantial differences in the provision of public housing to that in Britain and other Western countries. Public housing comprises about 3% of the housing stock in the U.S.A., compared with 30% in Britain. However, there is no serious doubt about Newman's theory that if housing design gives tenants a sense of having a personal stake in a piece of ground or part of a building and opportunity for surveillance of it this will deter crime.

One of his important points is that entrances in blocks of flats should lead only to a limited number of dwellings and be seen to be for residents alone. Through-ways to other locations and areas of semi-public space should be avoided, or at least be as restricted as much as possible and be open to surveillance.

With individual houses, where crime is likely to be restricted to burglary, a Home Office research and planning unit report (Winchester, S. and Jackson, J. (1982) *Residential Burglary: the Limits of Prevention*. London: H.M.S.O.) shows that the risk rises with environmental factors associated with the house which may be summarised as follows:

- The degree of separateness, e.g. detached contrasted with terraced.
- The less it is overlooked.
- The openness of the surroundings, e.g. bordering on a park or farmland where the burglar can make a getaway.

Where these factors are strongly in evidence a higher than normal grade of security should be considered because under favourable circumstances burglars are prepared to use as much force as necessary to effect an entry.

In 1983 Barry Poyner brought together in his book *Design Against Crime: Beyond Defensible Space* the fruits of the research that followed the publication of Newman's theories. From these studies Poyner defines thirty-one patterns of design for crime reduction. These patterns include presciptions of housing design that avoid the environmental factors described above and also avoid the 'openness' of main through-roads by facing houses onto side roads with access as direct as possible from these roads only. The need to restrict access from the rear of houses is also prescribed. For multi-unit apartment buildings the patterns include entrances that are either manned or 'kept locked with the potential of human surveillance through electronic means'. Another pattern requires apartment doors to be grouped around lobbies serving several apartments. Among the patterns for schools, an unobstructed view from the street and from any nearby houses is prescribed; where there is to be a resident caretaker compactness of design and a clear view of as much of the school as possible are dictated.

A determined criminal will get into the most secure building if he is given enough time. Fortunately his time is usually limited, the object therefore is to delay him as long as possible. Nothing must be made easy for him, whether it is working unobserved, climbing up to a window, forcing a lock, knocking a hole through a wall, or making an unsanctioned entry in any other way. Further delay is caused by defence in depth. Where possible, first the grounds or surroundings of the building are defended, then the building itself, then the area where the most attractive items are to be found.

There are advantages when designing for defence in depth by working away from the area where defence is required. By first identifying the targets (e.g. cash, goods, records), parts of the building that will house them can be grouped together as much as practicable. These can then be provided with more than basic security without the need for high-security construction at greater than normal cost over a considerable area, as would otherwise be necessary. This will be a **secure zone**. Beyond it in a **detection zone** basic security may be supplemented by intruder detection devices. Further out will be a **deterrent zone** where the intruder is deterred from entry by perimeter protection and lighting which will make unauthorised entry obvious to an observer. This approach to design can be particularly appropriate for the protection of records and items of value in school buildings where for economic and social reasons it is not possible to provide more than basic security for the main structure. For protection against nuisance and malicious attacks, reliance may have to be placed on deterrent and detection zones only.

If a building must be of light-weight construction, a security alarm system may be essential, in which case it should be designed in, although it may not be the ideal solution for the reasons explained in chapter 15.

Openings in perimeters and buildings themselves must be kept to a minimum. This applies to service ducts, air-conditioning vents, rooflights, etc., as well as to windows and doors. Any clear space in an opening through which an attacker might attempt to pass should not exceed 0.03 m^2 (0.33 ft^2) (see section 4.8).

For maximum security, windows below about 5 m (16 ft) should have bandit resistant glazing or be protected by steel grilles. If there is a risk of terrorism or rioting, ground floors may be windowless, though this medieval approach (the siege mentality) is criticised as leading to neglect of observation and control of the vicinity. The criminal may profit from this — one of the Great Train robbers escaped from prison by way of a pantechnicon which drew up against the prison wall. Anonymous blank walls with no surveillance are also an invitation to the graffiti artist.

As has been stated, certain populations of buildings may be judged as hostile but paradoxically, security measures have made vandals of responsible users. When access points are few, normally responsible people are likely to make 'illegal' short cuts for their own convenience. A medical consultant used an iron bar to force a way into a hospital through an emergency exit door; in his mind a consultant's convenience outweighed other considerations in such a building. This is an extreme case but it illustrates the need to keep in mind the effect of security measures on users of buildings.

Normally, vandalism will be reduced by:

- Ensuring, as far as practicable, that all parts of a building and its surroundings are overlooked or can be kept under surveillance.
- Making it obvious that someone has an interest in the area.
- Ensuring that the feature or object is not easy to damage and surfaces do not encourage graffiti.

3.6 Tamper resistant fasteners

As part of their general destructiveness, vandals may undo screws and bolts or loosen them so that handles or appliances come away when used — schools are particularly favoured by this trick. Thieves remove screws and bolts to gain entry. Defence against these attacks is given by fasteners such as those shown in fig. 3.3.

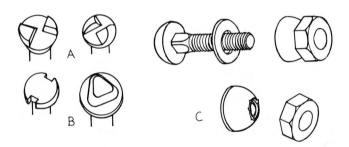

Figure 3.3 Tamper resistant fasteners. (A) Clutch-head or prison-head screws; (B) retractable screws requiring special tools for driving and extracting; (C) Orsogril anti-theft bolt, the nut breaks away when tightened to the correct torque.

What are known as either clutch-head or prison-head fasteners can be tightened by ordinary screwdrivers and spanners but cannot be removed by these tools, or indeed by any tools, without destroying the fastener or damaging the item secured. The retractable type of tamper resistant fastener is an alternative for use where removal by authorised personnel is necessary.

3.7 Assessing security requirements

Consideration of the following questions and typical answers about a building will assist determination of an appropriate grade of security and level of access control.

- Does it contain items of value to a criminal? (Easily disposed of items; costly, conveyable items; industrial or state secrets; incriminating information.)
- Are there users who are likely to commit crimes? (Persons likely to be tempted into opportunist crime or burglary; hot-tempered, frustrated persons.)
- Are there users who are likely to become terrorist targets? (Public figures; upholders of the law; ethnic minorities.)
- Does its use make it a terrorist target? (Government departments; political organisations; works of art; public records.)
- What is the attitude of users? (Hostile; responsible; security conscious.)
- Does its location increase the risk? (In large town or city, or within easy car journey; 'separateness' from other buildings.)

Chapter 4

Anthropometrics and Physical Performance

4.1 Fitting buildings to people

People show great versatility in adapting to buildings that have not been designed with their capabilities in mind. However if, in adapting, they have to over-exert themselves, stoop, over-reach or stand off the floor they are put at greater than necessary risk of having an accident. To minimize the risk, a designer must take into account the critical body sizes of users and the physical efforts they are capable of. The designer must consider not just the maximum physical capabilities of people, but how they are affected by ageing, the impairment of functions caused by temperature extremes and alcohol, and other debilitating influences. Additionally the physical limitations of children must be allowed for, as must their behaviour. For security, too, body size and physical capabilities must be considered, in this case primarily in relation to the building rather than the occupants.

4.2 The normal distribution, and percentiles

The principal difficulty in designing to fit people is due to their variability. Recorded adult heights range from 2 ft 2 in. to 8 ft 11 in. However, in spite of the physical variations, if stature or another bodily characteristic is measured in a large and representative sample of a population and the measurements are plotted on a graph (the dimensions on the horizontal axis and the frequency with which they occur on the vertical axis), as shown in fig. 4.1, a bell-shaped curve is obtained with the bulk of the sample clustered about the centre. The dimensions for the population as a whole are assumed to have a similar distribution, giving a curve that can be fitted by a specific mathematical formula. This is known as a normal distribution.

The fact that the bulk of the population is clustered about the centre of a normal distribution must not be taken as an invitation to design for the average person. This is not very beneficial. For one reason an individual of, say, average stature will most probably not be average in other dimensions; for another reason, some

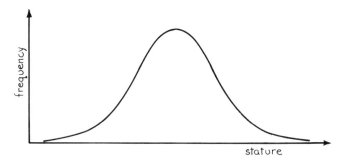

Figure 4.1 Normal distribution curve.

features based on an average value would be quite unsuitable for approximately half the population — if headroom were derived from average stature there would be a lot of sore heads!

The designer's aim must be to cater for as large a group of the population as practicable. Persons having extreme dimensions are not provided for, this would penalise the majority. Therefore when using anthropometric data it is usual to work from chosen percentile values. Percentiles are obtained by dividing the population into percentage categories, working from the smallest to the greatest value of the variable. Then the highest value in the smallest 5%, for example, is the fifth percentile of the population. Thus if the fifth percentile of stature for men is 1615 mm (5 ft 3½ in.) then 95% of men are taller. For some features it is appropriate to provide for 99% of users of buildings, headroom is such an example. For other features, various constraints on the designer will require him to make provisions based on 95 or 90%.

4.3 Regional and socio-economic differences

In different parts of the world the general build of the indigenous population differs. In addition to differences between one locality and another, there are differences between socio-economic classes and between the inhabitants of urban and rural areas. There are also ethnic differences: black Americans have longer legs than white Americans. Then there is the general increase in the size of people. In the developed countries, at least, children have been taller than their parents for several generations, though there are signs that this trend is now levelling out. At the present stage of anthropometric knowledge, data to allow for these secular changes and for regional and socio-economic differences are not available for building design. There are gaps even in data from limited samples and difficulties in reconciling data from one source with that of another. The designer can only use the best data available. Fortunately, most dimensions are not critical and a slight variation from the optimum will probably benefit as many at one end of the range as it deprives at the other.

4.4 Children

The first of two vital facts to remember when designing buildings that will ensure the safety of children is that children use buildings differently to adults. They lack experience to make accurate judgements about a building in the way that adults do. Adults make assumptions about the environment from very scanty information, two knobs about 0.5 m apart on a flat, vertical surface indicate a drawer, for example. A child sees the same knobs but may not interpret the information in the same way. Children are also more vigorous and exuberant users of buildings than adults, they run rather than walk and use parts of the building and its surroundings as play equipment (fig. 4.2). Table 4.1 shows the average ability of children, from nine months to five years old, to perform various activities of importance to safety design, and the precautions necessary. After the age of five years, climbing, sliding, swinging and the performance of stunts, such as jumping from one point to another, continue with a skill improved beyond that shown in fig. 4.3.

The other vital fact to remember is that, physically, children are not small replicas of adults. The proportions of their bodies differ considerably at different ages. The ratio of length of legs to total height varies from 1:3 in the newborn to 1:2 in the adult; at birth the ratio of vertical head height to total height is 1:4 whereas in adults it is 1:7½. The horizontal dimension of the head in small children exceeds the maximum body depth. This has led to accidents where a child has slipped feet first through an opening and has been caught by the head. For this reason the maximum spacing

Figure 4.2 A building can be a child's adventure playground.

Table 4.1 Children's ability up to 5 years of age* relative to safety design

Age	Development			Safety design
	General mobility and dexterity	Stairs activity	Climbing	
9 months	Progresses on floor by rolling or squirming; starts to crawl; pokes at small objects with index finger			Avoid gaps where any part of a child's body can become trapped; limit spacing of members of balustrades
12 months	Crawls on hands and knees; shuffles on buttocks; bear-walks rapidly about the floor	Crawls upstairs		
15 months	Walks alone with uneven steps; grasps pencil and imitates scribble	Creeps upstairs and gets downstairs backwards		Shield electric socket outlets
18 months	Runs carefully, but cannot, usually, continue round obstacles	Walks up and down stairs with helping hand; bumps down a few steps on buttocks, facing forward	Climbs into adult chair	Avoid climbing-stages in balustrades, standing-ledges by openable windows
	EXPLORES ENVIRONMENT ENERGETICALLY			
2 years	Runs on whole foot, stopping and starting with ease and avoiding obstacles; turns door knob	Walks up and down stairs holding on to rail or wall, two feet to a step	Climbs on furniture to look out of window, opens doors, operates switches	Provide children's hand-rail or handholds; avoid climbing-stages with bathroom appliances
2½ years	ACTIVE AND CURIOUS WITH LITTLE KNOWLEDGE OF COMMON DANGERS			
4 years	Navigates self-locomotion skilfully, turning sharp corners, running, pushing and pulling	Runs up and down stairs	Climbs ladders and trees	Avoid any feature capable of being climbed, squeezed through, slid down, swung from
5 years	ACTIVE AND SKILFUL IN CLIMBING, SLIDING, SWINGING AND VARIOUS STUNTS			

*Mainly from: Sheridan, M.D. (1975) *Children's developmental progress*. 3rd edn. Windsor: N.F.E.R. Publishing Company.

Figure 4.3 The sort of climbing ability possessed by a boy aged four to five years.

of bars in children's cots was fixed at 82.5 mm (3¼ in.) in BS 1753 (1965), however by the time the standard was revised in 1977 accidents of this type had occurred in Canada and the U.S.A. when the spacing was less than 82.5 mm. No similar accidents had been reported in the U.K. but it was felt prudent to reduce the maximum spacing to 60 mm (2⅜ in.). At an age when a child is crawling, a head dimension is still likely to exceed maximum body depth.

In safety barriers intended for the protection of children up to twenty-four months old, according to BS 4125 (1982) *Safety Requirements for Child Safety Barriers for Domestic Use*, bars are required to be spaced at 60—85 mm (2⅜—3⅜ in.). A diamond mesh is excluded and for a lattice-type mesh the maximum spacing permitted between adjacent vertical members is 25 mm (1 in.).

Children need to be able to reach light switches. Accidents have occurred where children have gone to the bathroom, been unable to reach the pull switch from the floor so have attempted to climb on to the bath to put the light on. The vertical grip reach given in fig. 4.4 is relevant to the height of switches, locks and other items.

From an early age a child explores its environment and, if it inserts its finger, hand, foot or head into a gap just big enough it may get the limb or head trapped. A well recognised example of this occurrence is a boy pushing his head through railings and having to be freed by firemen because his ears prevent withdrawal.

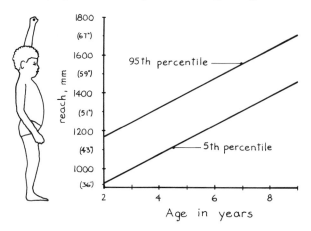

Figure 4.4 Vertical grip reach in children.

Anticipating this kind of happening, the compilers of BS 5696 Part 2 (1976), *Play Equipment Intended for Permanent Installation Outdoors*, in the matter of minimising hazards in play equipment, specified the use of probes (fig. 4.5) to investigate whether parts of play equipment accessible to a child's fingers, hand, foot, part of a limb, head or head and shoulders could form a trap. BS 3042 (1971) (confirmed 1980) *Standard Test Fingers and Probes for Checking Protection against Electrical, Mechanical and Thermal Hazard* is

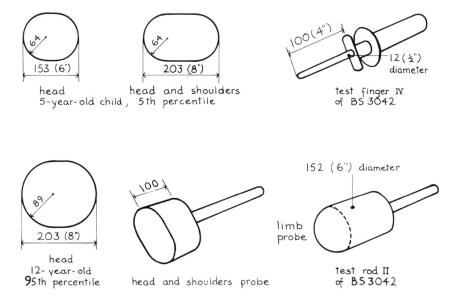

Figure 4.5 Probes to test children's accessibility to traps.

used when testing finger, hand, foot and limb hazards. Two head probes are specified in BS 5696, one simulating the head of the smallest likely user (fifth percentile five year-old) the other the head of the oldest intended user (ninety-fifth percentile twelve year-old). A head and shoulders probe simulates the smallest likely user. These and similar probes may have a use in completed buildings.

The building designer needs also to consider users smaller than five year-olds. Dimensions taken by Snyder *et al.* ((1977) 'Anthropometry of infants, children and youth to age 18 for product safety design.' Chicago: Society of Automotive Engineers) of approximately thirty children in each of the age groups 9—11, 12—15, 16—19, 20—23 months showed virtually no difference at the fifth percentile in middle finger diameter, minimum hand clearance and foot breadth through the whole age range, though crown—sole length increased from 662 mm (24½ in.) to 744 mm (29¼ in.). The dimensions are shown in fig. 4.6.

4.5 The elderly

With the greater life expectancy of modern times, the proportion of older people using buildings is steadily increasing. Soon 20% of the population of Britain will be over retirement age (sixty for women and sixty-five for men). The loss of acuity and range in the senses affects their safety. Most people need glasses to correct failure to focus at close range by the time they are fifty. Increasing age brings further sight deterioration and may necessitate the use of bifocal spectacles, which can lead to errors in judgement caused by looking through the wrong lenses. The time of reacting to a decrease in luminance is slowed by old age, thus impairing recovery from

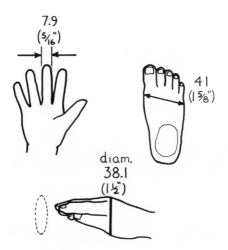

Figure 4.6 Finger, hand and foot dimensions (in mm), fifth percentile 9—23 months.

glare. Disturbance of vision is also caused by glaucoma — abnormally high pressure within the eye — to which older people are prone. By the age of seventy, vision is dimmed to some degree in over 90% of people by senile cataract. What the elderly *can* see they often have difficulty in interpreting; an incorrect or incomplete visual interpretation made on the spur of the moment will often put their safety in jeopardy.

Slowness to recover after being thrown off balance, coupled with poor posture, contributes to falls in the elderly. Dizziness occurs through the momentary lowering of blood pressure. The bones of elderly people are brittle and thus it is common for them to break in a fall which would be almost unnoticed in a child or young person. Dropping only a small distance can be dangerous. Even the strong bones of healthy young men sometimes break after a very short fall. The voluntary reponse to starting to fall takes over 190 milliseconds, corresponding to a fall of about 180 mm (7 in.), the reflex response acts faster; nonetheless, a landing after a fall of less than 180 mm will be largely uncushioned by muscular action.

Old people are shorter than young people. This is due partly to compression of the lumbar discs and slackness of the postural muscles, and partly to the lesser stature of the generation to which the old people belong. In matters of reach, therefore, if an item is within the compass of an elderly woman (fig. 4.7) its position is likely to suit the rest of the adult population. Hence most of the critical dimensions for safety shown in table 4.2 are derived from the body sizes of elderly women.

Figure 4.7 Elderly woman's limited ability to reach and stoop (see also table 4.2).

Table 4.2 Critical dimensions for safety

Reference	Bodily Characteristics: Description	Dimensions mm	Dimensions in.	Percentile	Applications: Description	Dimensions: Mode	Dimensions mm	Dimensions in.
	Elderly women, with low heels					*Height above floor*		
Section 4.5	Eye height, standing*	1330	52½	5th	Window sill	Max.	1220	48
					Vision panel, bottom edge (but see section 12.3)			
					Shelf, seeing into	Max.	1320	52
					Meter for reading	Max.	1450	57
	Shoulder height, standing*	1180	46½	5th	Controls, e.g. thermostat	Max.	1200	47
	Elbow height, standing*	880	34½	5th	Safety grasping rail in corridor	Pref'd	910	36
Fig. 4.7	Maximum upward reach to finger/thumb grasp*	1590	62½	1st	Emergency switch for standing use, e.g. cord operated			
					Door bolt	Max.	1570	62
					Window fastener with no obstruction in front of window			
	Ditto with 360 mm (14 in.) obstruction in front of body*	1440	56½	1st	Window fastener with obstruction in front	Max.	1420	56
					Shelf, without vision			
	Ditto with 600 mm (2 ft) obstruction 900 mm (3 ft) from floor†	—	—	1st	Window fastener with obstruction in front	Max.	1330	52
Fig. 4.7	Height from floor to fist when bending trunk forward and downward*	290	11½	(Average)	Electric socket outlet	Min.	300	12
					Lowest shelf			
	Men, age 25—34, with 25 mm (1 in.) heels							
Section 4.6	Stature**	1955	77	99th	Headroom	Min.	2000	79
	Centre of gravity (57% of stature)**	1125	44¼	99th	Balcony railings	Min.	1150	45¼
Section 4.6	*Men, all ages*					*Width of tread*		
	Foot length, from heel to ball of foot**	214	8½	95th	Tread of step	Rec'd	298	11¾
	Foot length, without shoe**	291	11½	95th				
	with normal shoe**	328	12 15/16	95th				
	with heavy winter boots**	363	14 5/16	95th				
	Men, flyers[ξ] *with 25 mm (1 in.) heels*					*Below elbow height*		
Section 4.7	Elbow height**	1226	48¼	95th	Working surface (general)	Pref'd	75—200	3—8
		1074	42¾	5th				
	Women, all ages, with shoes							
	Elbow height‡	1095	43	95th	Working surface (kitchens)	Pref'd	90—120	3½—4¾
		935	36¾	5th				

*BS 4467 (1969) *Anthropometric and Ergonomic Recommendations for Dimensions in Designing for the Elderly*, samples taken in Oxford region and Birmingham.
†Covington, S.A. (1982) 'Ergonomic requirements for building components and associated operating devices'. *B.R.E. Current Paper 1/82* Garston: B.R.E. (Subjects from Loughborough).
**Panero, J. and Zelnik, M. (1979) *Human Dimension and Interior Space*. New York: Whitney/London: Architectural Press (USA data).
ξU.S. Air Force flying personnel.
‡Ward, J.S. and Kirk, N.S. (1970) *Ergonomics*, **13** (6), 783—797.

4.6 Large men

Headroom, as previously mentioned, needs to be based on the stature of the ninety-ninth percentile man. With steps and stairways, foot length at the ninety-fifth percentile (table 4.2) may be the basis of design. In deciding the height of balustrades and barriers provided for safety, the position of the body's standing centre of gravity in the taller person is of importance. In the upright position, with the arms hanging by the sides, the body's centre of gravity is in the pelvic region, usually at 55—57% of total stature in adult males, reckoned from the soles of the feet. At the ninety-ninth percentile of stature, 57% puts the standing centre of gravity at 1100 mm (3 ft 7¼ in.) from the floor, 1125 mm (3 ft 8¼ in.) with shoes: a dimension that is consistent with other available data. (Snyder *et al.* obtained 1064 mm, from soles of feet, for the ninety-fifth percentile of males and females 17.5—19.0 years old.)

4.7 Impairment of performance

If a person's performance of a task is affected by environmental conditions then mistakes leading to accidents are likely. Fatigue is one of the factors to guard against. Standing at a bench or worktop demands correct posture to avoid fatigue in the short term and progressive physical degeneration in the long term. The human back is supported best by its muscles and ligaments when it is upright. Bending forward over a work surface puts support for the trunk on the ligaments of the vertebrae and makes little use of muscles. A ligament may be stretched or an intervertebral disc injured by the strain, resulting in back pain. On the other hand too high a work surface, leading to raising of the shoulders to position the hands above the work surface, requires a relatively steady contraction of muscles not directly involved in the work, this causes pain and muscle weakness and is 'tiring'.

The critical dimension for comfortable and efficient work when standing is elbow height. Work surfaces need to be sufficiently below elbow height to allow tools and utensils to be handled without hunching the shoulders or raising the upper arm sideways, but the surfaces must not be so low as to cause stooping. Various studies have shown preferred heights for working surfaces from 75 mm (3 in.) to 200 mm (8 in.) below elbow height, according to the work done and the size of the tool, utensil or apparatus handled.

For kitchen working surfaces Ward and Kirk (see table 4.2) experimentally determined preferred heights, below elbow height, of 88 mm (3½ in.) for light work and 122 mm (4¾ in.) for work requiring force on the surface. Standard kitchen fittings with fixed worktop heights cannot be suitable for more than a small proportion of the population, hence it is desirable that heights should be adjustable. Figure 4.8 shows one way in which this can be accomplished.

Figure 4.8 A wall mounted kitchen unit with height adjustment (Phlexiplan).

Discomfort caused by too hot an environment impairs performance, so also does clumsiness caused by too cold an environment. Deterioration in performance leading to accidents has been found to occur when air temperature rises to 27°C (80°F) or falls to 13°C (55°F). Dim lighting is an obvious danger. Too much light in the form of glare can also be dangerous (see section 9.2).

The effect of the consumption of alcohol on driving skill and the relationship with road accidents is well recognised. Less attention has been given to the influence of alcohol on other forms of accident. Where it has been investigated, as in the study of fatal home accidents mentioned in section 2.2, it has been found to be of greater extent than had been previously recognised and not adequately taken into account when considering preventive measures. Probably the same can be said about the misuse of drugs. Even a double dose of sleeping tablets (200 milligrams of quinalbarbitone sodium) will depress brain function sufficiently to produce a reliable deterioration in performance ten hours later. The increase in social drug taking may be reflected in an increase in accidents including those where building design plays a part by permitting the escalation of an error of judgement into an event causing injury or death.

4.8 The intruder

As most vandalism and most burglaries are committed by young people, it is the body sizes of young people (fig. 4.9) that the designer needs to know when designing to keep out the vandal and burglar. The question of age is relevant in this respect. Children below ten years old, the age at which they can be held responsible in law, have been responsible for a considerable amount of vandalism, and they have been used by adults to get through small openings to give them entry to a building. Thus a ten year-old may be taken as the youngest lawbreaker against whom precautions need to be taken and the maximum size of openings permitted may be based on body sizes of a child of this age. Hand and arm dimensions, relevant to the size of opening through which the upper limb can be passed, are given in table 4.3. Teenagers are a greater danger to the security of a building, so dimensions of their age groups are also given.

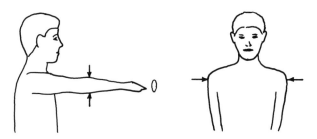

Figure 4.9 Some critical body sizes of a potential vandal and burglar (see also table 4.3).

Table 4.3 Some critical body dimensions for security (fifth percentile)

Age in years	Males		Females	
	mm	in.	mm	in.
9.5—10.5				
Minimum hand clearance	51	2	48	1⅞
Forearm breadth	51	2	50	2
Shoulder breadth	298	11¾	286	11¼
13.5—14.5				
Minimum hand clearance	55	2³⁄₁₆	57	2¼
Forearm breadth	64	2½	62	2
Shoulder breadth	348	13¹¹⁄₁₆	340	13⅜
17.5—19.0				
Minimum hand clearance	67	2⅝	—	—
Forearm breadth	73	2⅞	60	2⅜
Shoulder breadth	405	15¹⁵⁄₁₆	368	14½

From Snyder, R.G., Spencer, M., Owings, C. and Schneider, L. (1977)
Anthropometry of infants, children and youth to age 18 for product safety
design. Final report. USA Society of Automotive Engineers, to the nearest mm.

Shoulder width is included in table 4.3 for guidance on the size of
opening a body can pass through; unfortunately other useful
dimensions, such as body depth, are not available. However it seems
likely that a ten year-old would be able to squeeze through an
opening of 280 x 125 mm (11 x 5 in.), i.e. an area of 35 000 mm²
(55 in.²), hence it is advisable that no clear opening should exceed
30 000 mm² (47 in.²).

4.9 Checklist

Safety
Make provision for:

- Children's
    playful and exploratory activities (table 4.1)
    limited reach (fig. 4.4)
    not trapping body parts (table 4.1, fig. 4.5)
- Elderly persons'
    defective vision
    infirmities
    limited eye height, reach and stoop (table 4.2)
- Large men's
    stature (table 4.2)
    feet (table 4.2)
- All users'
    fatigue from unsuitable height of working surface (table 4.2)
    impaired performance and irrational behaviour due to unsatisfactory environmental conditions, alcohol, drugs and other factors

Security
Determine maximum sizes of permitted gaps from body dimensions of ten year-old children and teenagers (table 4.4).

Chapter 5

Health Hazards

5.1 General causes

Buildings give rise to health hazards firstly because materials from which they are built and the equipment they contain emit pollutants and secondly because the buildings prevent dispersal of these pollutants and of other pollutants introduced by occupants or outside agencies. Minor effects on health are: headaches, stuffiness, eye, nose and throat irritation. Serious effects are: respiratory infection, increased risk of cancer, and aggravation of blood disorders and heart disease.

Houses in Britain are now a more closed environment than twenty-five years ago. They are 'tighter' mainly because of draught stripping of external doors and windows, and because of the removal of open fires. Other buildings are tighter because of improved draught stripping and other energy-saving measures. Reliance on natural ventilation in these tight buildings leads to increased contamination of the air from the sources given in table 5.1 and from pollutants introduced by occupants, such as tobacco smoke, animal dander, and solvents in cosmetics, cleaning materials and hobby materials. On the other hand, the air that is drawn in from outside is generally less contaminated than it was before the implementation of the Clean Air Act 1956. Since the passing of the Act there has been a steady reduction in the amount of smoke and sulphur dioxide in the air over Britain.

Now that it is known that 'passive' or 'secondary' smoking carries an increased risk of lung cancer, the most harmful pollutant of which we are aware, which is introduced into non-industrial buildings by occupants, is likely to be tobacco smoke. To dilute the 'side-stream' of one cigarette, about $20 \, m^3$ ($700 \, ft^3$) of fresh air are needed. The average smoker smokes 1.3 cigarettes an hour, so $26 \, m^3 \, h^{-1}$ of fresh air are required for every smoker. On this basis one air change per hour (a.c.h.) is necessary for a room 4.75 m (15 ft 6 in.) square and 2.3 m (7 ft 6 in.) high occupied by two smokers. In a flueless room of this size, in a brick built building with normal windows and door without weatherstripping, the a.c.h. would be below 0.5. If reliance is to be placed on natural ventilation

Table 5.1 Pollutants resulting from building planning, materials or services

Source	Pollutant	Health effect	Recommendation
Uranium enriched: ground soils, granite, ground water, concrete aggregate, plasterboard	Radon and progeny	Increased risk of cancer	Seal over site Provide ventilation under the ground floor Increase room ventilation when heating is at low level Avoid use of contaminated material
Vehicles in underground and attached garages, and on main roads Unventilated heaters Gas cookers Faulty heating system	Carbon monoxide, CO Carbon dioxide, CO_2 Nitrogen dioxide NO_2	Aggravation of blood disorders and heart disease Interference with respiratory functions	Extractor fans for garages No residential buildings near busy traffic junctions Flues for heaters Extractor fans for kitchens
Urea formaldehyde in: cavity wall insulation, chip(particle) board, laminated wood products	Formaldehyde	Respiratory tract irritation	Control of cavity insulation Increased ventilation for period after installation
Lead water pipes Lead-based solder in copper piping Vehicles Gloss paints containing lead	Lead	Lead poisoning	No use of lead pipes W.C. flushing cisterns at end of pipe runs Lead-free solders Vehicles (as above) Paints with lead content not exceeding 0.06%
Asbestos products	Asbestos fibre	Asbestosis Cancer	Do not use
Miscellaneous: adhesive solvents, fungicides in paints, interior decoration, mould and fungal spores arising from damp conditions	Various	Eye and respiratory tract irritation Allergies	Increase room ventilation during, and for period after, use Inhibit growth of moulds and fungi by control of rising damp and humidity

in an office or living room of this size it would, therefore, be advisable to provide a non-closable wall ventilator.

5.2 Radon

Radon is an inert gas resulting from the decay of radium; it is short lived and decays to other radioactive elements. Inhalation of radon and its progeny brings radioactivity into the lungs, where irradiation of tissue can lead to cancer. Smoking increases the risk of the disease gaining a hold.

The radium which produces cancer is itself a decay product. It evolves from uranium, and uranium is a trace constituent of soils and rocks. Buildings contain radon derived from the ground on which they stand, from the 'earthy' materials from which they are constructed, and from the outside air. In a typical house in the United Kingdom about 50% of the radon present comes from the ground, about 25% from building materials and about 25% from the outside air. Granite and some shales and sandstones have higher than average concentrations of uranium, it is therefore important to be aware of the problem when building on ground where these are present. In the U.S.A. concern is felt about the risk arising from building on land reclaimed from phosphate mining. The nature of the building materials is less important than the ground, but granite produces about twice as much radon as clay bricks. In certain materials the manufacturing process may result in a much higher concentration of radioactivity than in natural materials, an example is plasterboard made from by-product gypsum from the phosphate mining industry.

A low ventilation rate in a building allows the concentration of radon to rise and thus the risk to health is increased. Lowering the concentration by increasing ventilation is unfortunately at variance with the need to conserve fuel by reducing heat loss. Alternative solutions, on which research is proceeding, are increasing the sub-floor ventilation and the use of barriers and sealants to prevent the gas diffusing into the building. High tensile polyamide film and epoxy sealants have been used to give over-site protection. For existing buildings the solution seems to lie in arranging for generous ventilation at times when this does not cause a substantial heat loss. An alternative solution, applicable to new and existing buildings, is to use mechanical ventilation in conjunction with a heat exchanger. Heat extracted from expelled air is used to warm fresh air as it is pumped in.

The Royal Commission on Environmental Pollution says in its tenth report that it would be prudent to design new buildings to be within a dose limit of 5 mSv per year (the sievert (Sv) is a measure of risk for all forms of radiation of the body). However the Royal Commission acknowledges that there is difficulty in identifying problem sites.

5.3 Oxides of nitrogen and carbon

Nitrogen dioxide (NO_2) is a product of high temperature combustion; carbon monoxide (CO) and the relatively harmless carbon dioxide (CO_2) are products of low temperature combustion.

Nitrogen dioxide
Sources of nitrogen dioxide in building are combustion appliances and motor vehicle exhausts. It is difficult to determine the effect of low concentrations on health because of the presence of other contaminants in indoor air. For example, an association has been found between the prevalence of respiratory illnesses in primary school children and the use of gas for cooking in the home. This association is thought to be due to combustion products of unflued gas cookers, in particular nitrogen dioxide, but in the absence of direct proof the researchers are unwilling to come to a firm conclusion about the effects — if there are any they are likely to be slight.

Carbon monoxide
Carbon monoxide is an almost odourless gas. It is very poisonous. When inhaled it combines with haemoglobin, the chief oxygen carrying substance in the blood, about 250 times more readily than does oxygen. Carboxyhaemoglobin is formed and thus haemoglobin is no longer available to carry oxygen to body tissues. A concentration of 0.5% of carbon monoxide in the air will cause rapid collapse, unconsciousness, and death within a few minutes. At low concentrations the symptoms include increased pulse and breathing rate, lassitude, nausea and headaches. Naturally, children and people with lung, heart or circulatory disease are more susceptible than healthy adults.

Deaths attributed to carbon monoxide poisoning from domestic fuels in the U.K. totalled 130—160 in 1979, over 100 being caused by gas appliances. Later estimates put the figure for deaths from carbon monoxide poisoning in the home at nearly 200 a year. Blockage of flues of gas fires installed in fireplaces formerly used for coal burning is one of the causes of these deaths. If the gas fires are improperly installed, falls of soot can block the flue outlet at the back of the appliance, allowing carbon monoxide to build up undetected in the room. Deaths have also resulted when flues have been blocked by the escape of cavity insulation foam.

Carbon dioxide
In an unventilated room shortage of oxygen for breathing is unlikely to be a threat to the occupants, the accumulation of carbon dioxide, either exhaled or from the combustion of fuels, is more serious. Country air contains only about 0.3% carbon dioxide, but a concentration of 4% is tolerable, though unpleasant. Even when carbon dioxide threatens life at 6% concentration, the oxygen

content will not have fallen to a dangerous level. While it is most unlikely that a building would be sufficiently airtight for enough carbon dioxide to accumulate to endanger life directly, an excess of carbon dioxide in the air supply to gas and oil heaters can affect their performance, leading to incomplete combustion of the fuel and excessive formation of carbon monoxide.

An adequate air supply for heating appliances provided with flues is usually met by a permanent ventilation opening of area equivalent to that of the flue connection. For a flueless appliance such as a domestic gas cooker, an opening 9000 mm² (15½ in.²) in area is usually sufficient.

5.4 Formaldehyde

Formaldehyde is a gas which is very soluble in water. In buildings it is used mainly in cavity wall insulation and in the manufacture of wood chipboard and other particle boards. Formaldehyde is also used in many everyday items found in buildings, such as carpets, furniture and fabrics. It is found in small quantities in clothing and is emitted by gas cookers and by smoking. It is also produced by the human body.

In the report *Indoor Pollutants*, by W. Ashington of the U.S.A. National Research Council, published in 1982, it is suggested that perhaps 10—20% of the general population may be susceptible to the irritant properties of formaldehyde at extremely low concentrations. Symptoms reported by people exposed to the fumes embrace headaches, sore throats, smarting eyes, coughs, sneezing, shortness of breath, nose bleeds, severe skin irritation, eczema-like rashes, nausea and aching limbs. Formaldehyde was found to be a carcinogen causing nasal cancer in rats when they were exposed to the gas. This experiment caused widespread concern when its results were known, however the rats had been exposed to concentrations of formaldehyde higher than its pungent smell would permit humans to tolerate. For this reason, and because an increased risk of nasal cancer has not been found in people working with formaldehyde, it is not generally believed that the results are applicable to humans.

When formaldehyde is used for the insulation of cavity walls of existing buildings, a hardener and a surfactant are added to an aqueous solution of urea-formaldehyde. The surfactant lowers the surface tension of the liquid and enables it to form a foam. This is injected through holes bored at intervals in the outer leaf. In a short time the foam sets, then it dries out by evaporation over a longer period. Generally it is inappropriate to use this method of insulation other than in walls of sound and continuous double masonry construction, otherwise gas released from the foam will enter the building through cracks and breaks in the inner cladding. In Britain the use of urea-formaldehyde foam insulation in walls having masonry inner and outer leaves is governed by a code of practice, BS 5618 (1978).

Fumes from formaldehyde used as a bonding agent in manufactured

boards may be emitted for a considerable time after installation.
BS 1142 *Fibre Building Boards*, Part 2 (1971), Part 3 (1972), and
BS 5669 (1979) *Wood Chipboard and Methods of Test for Particle
Board*, put limits on the emission characteristics of these boards,
but the specified limits may be exceeded by boards not conforming
to the British Standards.

5.5 Lead

Like formaldehyde, lead is everywhere; it has been used and
dispersed by man for thousands of years. No other toxin is known
to be so widely distributed in the general population. Most of the
uptake is from food and drink, about 20% is from petrol. Lead
accumulates in the body, mostly in the bones. High concentrations
cause lead poisoning with definite signs and symptoms, low
concentrations are thought to affect the intelligence, behaviour
and performance of children, though there is considerable
uncertainty about this.

Buildings contribute to lead in the body through the use of
lead in the water supply and lead in paint. Soft water dissolves
lead more readily than hard water. Lead plumbing is not now used
but lead solder is used for joining copper pipes and poor workmanship
can result in the lead coming into contact with the water and entering
it through electrolytic action, particularly in soft water districts.
Where water is not drawn frequently, the lead content may rise to
unacceptable levels. The problem can be avoided at a small extra
cost by the use of lead-free solders.

Paints with white and red lead pigments were once in common
use and they are still causing lead poisoning in children who suck
and chew them. Today these paints are recommended only for
exterior use; even there they can be dangerous when the time comes
to remove them. Some oil-based paints contain organic lead as a
drier, though alternatives are available. When the lead content is
more than 0.5%, paints and varnishes are required by E.E.C.
regulations to carry a warning notice. In the U.S.A., the Lead-based
Paint Poisoning Prevention Act places a limit of 0.06% on the level
of soluble lead in paint accessible to children. The Royal Commission,
in its ninth report, recommended that Britain should work towards
reducing lead in paint to the U.S.A. level and that, in the meantime,
all paint containing more should carry a warning label.

5.6 Asbestos

Widespread concern about the danger of asbestosis and cancer of
the lung caused by the inhalation of asbestos fibre has led to a
decline in the use of asbestos products in building. Import of the
more dangerous blue (crocidolite) and brown (amosite) types into
the U.K., and the manufacture and supply of products containing
them, is now banned. The use of the white (chrysolite) type in
building is confined mainly to asbestos-cement products, with

some use in mastics and textured paints. Asbestos-cement contains 10—15% asbestos and the fibres are firmly bound in the cement matrix. Provided the products do not become damaged, or deteriorate with age, the fibres will not be released to the air.

Asbestos-cement weathers when used externally but the release of fibres is slow enough not to give rise to a significant risk to health. Even this can be prevented by sealing with a paint or plastics protective film. For interior use, sealing is generally essential because any fibres released may accumulate. Though under these conditions the health risk to the general public is insignificant, fibres are likely to be released when the product is worked or damaged during installation or removal, putting operatives and perhaps the public at risk. The use of asbestos in any form is therefore best avoided if a satisfactory substitute is available.

Less is known about the health hazards of mineral fibre substitutes than about asbestos itself. There is some evidence of risk, but at a much lower level than that from asbestos. Care should be exercised in the installation of substitutes and they should be protected from accidental damage and vandalism. (See section 10.7.)

5.7 Other health risks

Catching an infectious disease in the home is not usually regarded as preventable by building design in the way that injury or pollutant risk is preventable. In the case of 'tummy bug', however, it may be. The major cause of the spread of gastro-intestinal infection through families is the lack of adequate handwashing. Every W.C. should be complemented by a washbasin in the same room. It is not sufficient to rely on the use of an adjoining bathroom — it might be occupied when needed.

Legionnaires disease is a form of pneumonia caused by a bacterium, *Legionella pneumophila*, that occurs in nature under widely varying conditions and may be found in the water in air conditioning and other recirculating cooling systems. A prime means of infection is thought to be inhalation of a fine spray of water containing the bacteria. Planned maintenance to keep systems clean and inhibit the presence of bacteria is essential, additionally the water may be treated with a biocide effective against *Legionella pneumophila* (e.g. Houseman Hatacide LP5).

Regular cleaning of recirculating water systems of humidifiers and chlorination of the water will inhibit the growth of micro-organisms that cause humidifier fever.

Chapter 6

Glass, and Alternative Materials

6.1 Annealed glass

Probably the most dangerous and vulnerable material that will ever be used extensively in buildings is annealed glass. This is the type of glass normally used for glazing, and were it discovered today for the first time it would undoubtedly be considered too dangerous and too vulnerable to use. Most annealed clear glass now used in buildings is of the float glass type. In its manufacture the glass from the furnace is floated across a bath of molten metal. Sheet glass is made by an older, drawn process and is optically less acceptable. These glasses are dangerous and vulnerable because a pane of annealed glass does not plastically deform to any significant extent when struck. Its failure to dent means that the forces of impact are spread over only a small area, producing high stresses in the material. Cracks form and are propagated, splintering the pane into sharp dagger-like shards. When the glass is of the patterned type (cast glass) it is even more readily splintered.

There are safer and more secure alternatives to annealed glass, unfortunately their initial cost is greater. However in many situations the use of annealed glass does not make good economic sense. Leaving aside the cost of human misery caused by accidents, there is a high price to pay for medical treatment and replacement of materials. Vandalism can lead to deterioration of the building and its contents, losses from burglary are likely to be many times greater than the cost of building material damage.

Each year some 32 000 injuries treated in hospital and emergency departments in England and Wales are caused by glass which is in situ and not being repaired or replaced. This figure may be broken down to 20 000 injuries caused by glass in doors and 12 000 by glass in windows. About 47% of these are injuries to children under fifteen years of age. To these figures must be added those from accidents involving non-domestic doors and windows, glass in cupboards, kitchen fitments, shower screens and other fixtures.

The U.S. Consumer Product Safety Commission estimates that more than 180 000 architectural glass injuries are treated annually in hospitals in the U.S.A. Ordinary windows, excluding storm

windows and jalousie (louvred) windows, are associated with
94 000 injuries and doors with 73 000 injuries.

As would be expected, most of the injuries are cuts and lacerations.
Deep cuts sever tendons and nerves, and thus may permanently
affect motor functions. Deaths occur from the cutting of main
arteries. The distribution of parts of the body injured as given by
Clark and Webber ('Accidents involving glass in domestic doors
and windows: some implications for design.' *B.R.E. Information
Paper IP 18/81.* Garston: B.R.E.) are shown in fig. 6.1. As it is
most likely that an adult victim's arm was below shoulder height
when it was injured, fig. 6.1 confirms what we might have expected,
that is, that the greatest area of risk even to an adult lies below
1500 mm (5 ft) from floor level; the danger to a child is explicit.
Thus very careful consideration should be given to the use of
annealed glass in panels where any part is below the 1500 mm
level. This requirement is emphasised by the examples given in
chapter 12, for doors, and chapter 14, for windows.

The fragility of annealed glass makes it the material most
vulnerable to vandalism and housebreaking (in the literal sense
of the word) and the noise it makes on breaking and falling to the
ground gives the vandal added satisfaction. To the less-determined
burglar, the noise of breaking and falling glass may be a deterrent.
Not so to the experienced criminal; an old trick used to be to smear
treacle on to brown paper and stick it on the glass to hold the
broken pieces together — nowadays the same effect can be
accomplished with stout, sticky tape. Some other ways of breaking
in through glass are shown in fig. 6.2.

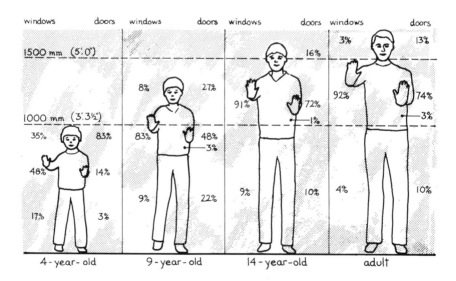

Figure 6.1 Males of fiftieth percentile height, and distribution of parts of the
body injured by glazing in age group 0—4, 5—9, 10—14 years, adult: head,
upper limb, trunk, lower limb.

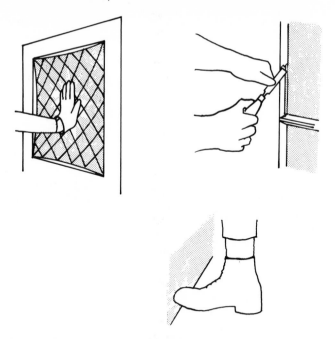

Figure 6.2 The criminal's way with glass: pushing in a leaded light, removing new putties, booting-in low-level glazing.

As with safety glazing, materials for security glazing are considered here in a general way. Specific uses are dealt with in chapters 8, 12, and 14.

6.2 Safety and security requirements of glazing sheet materials

Nowhere else in building design are the requirements of safety and security so closely allied as in the selection of a glazing sheet material. Of course the relative importance of safety and of security will depend on the location of the glazed areas, but always the selection of the material must start from the determination of which of the following safety and security functions the material needs to fulfil:

- Not inflict injury to people making accidental impact.
- Not break in a way that will significantly increase risk of injury from an explosion.
- Resist accidental penetration by a human body.
- Provide a barrier to the unauthorised passage of people into or out of a building.
- Resist attack by vandals.
- Resist attack by determined criminals.
- Resist attack by armed criminals.
- Resist blast waves from explosions.
- Resist the passage of fire.

When a glazing sheet material is broken, accidentally or deliberately, the forces involved are usually concentrated and short acting. Knowledge of how glass and alternative materials behave when subject to such forces is of prime importance for safety and security, therefore tests have been developed to examine this behaviour and give guidance in the selection of a suitable material. Most of the tests are of the dropping ball type. There is little standardisation in the tests of various agencies, the mass and substance of the ball differ, so does the size of the test piece. Comparative results on different materials are obtained but unfortunately it is difficult to relate the results of one form of test to another. If a standard test were available with only one variable, e.g. the minimum distance the ball must drop to cause breakage of the material, we would then be able to understand in a simple way what was involved in claims that a material is so many times stronger than glass. The British Standard tests described in section 6.3 do not provide this information, they are of the 'go, no-go' type; the test piece either passes or fails.

6.3 Impact tests for glazing sheet materials

Safety test
Impact performance requirements for flat safety glass and safety plastics for use in buildings are specified in BS 6206 (1981). The method of test is based on that developed by the American National Standards Institute (ANSI Z 97.1 (1975)). Three classes of safety material are determined by the method shown in fig. 6.3.

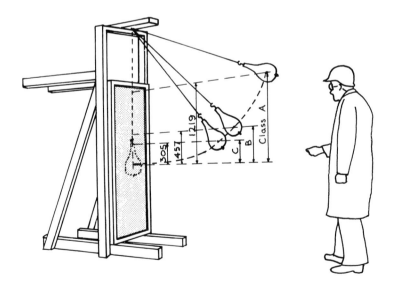

Figure 6.3 Drop heights (in mm), for classifying safety glass and safety plastics by impact test (drop heights in feet are 4, 1.5 and 1).

Test pieces have to be 865 ± 3 mm wide by 1930 ± 3 mm high, or the maximum size available if that is smaller. A leather bag (the impactor) filled with lead shot to weigh 45 kg (100 lb) is released from the heights shown to swing against the test piece. For a material to pass the test it must either NOT BREAK or BREAK SAFELY. If four test pieces pass when struck by the impactor from all three of the specified drop heights it is a class A material; if they pass from the two lower heights only it is a class B material; if they pass from the lower drop height only it is a class C material.

Ways that the material may break safely are defined in BS 6206 as follows:

(a) Numerous cracks or fissures appear but no shear or opening develops through which a 76 mm (3 in.) diameter sphere could be passed freely.
(b) Disintegration occurs but the ten largest crack-free particles remaining three minutes after impact together weigh no more than the mass equivalent to 6500 mm² (10 in.²) of the original test piece.
(c) Breakage results in several separate pieces but none of these presents sharp edges which are pointed or dagger-like. Pieces are not considered separate if they are retained on an interlayer or surface film or sheet.

The standard requires panels of safety glass and safety plastics to be marked with the manufacturer's name or trade mark or a mark that will identify the company who last cut the material if the original manufacturer's name is removed during cutting. Also required are the number of the standard, BS 6206; the classification of the material, A, B or C; the type of material, e.g. 'L' for laminated, 'T' for toughened, 'P' for plastics.

The impactor simulates a child weighing 100 lb (about average for a twelve year-old). Such a child running full pelt would develop kinetic energy in the order of twice that of the impactor dropping from 1219 mm, but the child's full energy would not be transmitted on contact with a glazed area. Perhaps first the hands, then the head, then the knees would make contact, and the incident angle of their impact with the surface would seldom be at right angles. Dropping the impactor from the three specified heights gives it impact energies of 135 J, 202 J and 538 J respectively. These are considered representative of impact energies likely to be delivered accidentally by persons to glazed areas. Thus the designer has to anticipate the kind of accidental contact with a glazed area he has to guard against. What energy is likely to be developed? Is it sufficient just for the glazing to break safely, or must it still provide containment after it has broken?

Some guidance to the grade of glazing necessary might be obtained by assessing the speed at which the glazing would be struck. If a child weighing 45 kg (100 lb), is assumed to develop values of kinetic energy on running twice those of the tests, then because the kinetic energy of any body at any instant is equal to ½ mass x (velocity)^2, the velocity of the child corresponding to, say, the A grade is evaluated as follows:

$$\text{velocity} = \sqrt{\frac{2 \times (2 \times 538)}{45}} = 6.9 \text{ metres per second.}$$

The three velocities corresponding to the glazing grades are thus:

A	6.9 metres per second	(15.5 m.p.h)
B	4.2 metres per second	(9.4 m.p.h.)
C	3.5 metres per second	(7.8 m.p.h.)

For an adult weighing twice as much (i.e. 90 kg) the velocities to develop the same values of kinetic energy would be $\sqrt{½}$ or 0.7 times as much. Class C glazing would most probably withstand a person of this weight walking into it but could not be relied upon to withstand him striking it bodily at a speed of 5.5 miles per hour. However with an adult the question is not likely to be whether he is running, but whether he is falling and how he might strike the glazing. As a general rule, class C glazing is the minimum standard for low level glazing and class B for glazed doors and side panels, though consideration should always be given to the need for a higher grade.

Test for bandit resistance
Two tests are specified in BS 5544 (1978) *Anti-Bandit Glazing*, and in ANSI UL 972 (1978) *Burglar Resistant Glazing Material*, for resistance to manual attack, as might be carried out with a pickaxe handle, a sledge-hammer or an iron bar. One test is to determine resistance to successive impacts of low energy, the other to determine resistance to a high energy attack. They are dropping ball tests carried out with apparatus of the kind shown in fig. 6.4. The ball is of hardened steel with a mass of 2260 gm (5 lb) and a diameter of approximately 82 mm (3¼ in.). It is released by an electromagnet to fall on to a test piece clamped between two steel frames fitted with rubber gaskets. The exposed area of the test piece must be at least 550 mm by 550 mm (1 ft 9⅝ in.). To pass the successive impacts of the low energy test, the test piece must survive five impacts from a height of 3 m (10 ft for ANSI UL 972) without the ball penetrating. To pass the high energy impact test, the ball must not penetrate when dropped from a height of 9 m (40 ft for ANSI UL 972). Test pieces are also subjected to a boil test for an indication of their durability. It is also a requirement of BS 5544 that glazing panels supplied as conforming to the standard are indelibly marked to indicate the fact.

Test for bullet resistance
Because bullets from different guns differ in their penetrating power, BS 5051 *Security Glazing* classifies bullet-resistant glazing according to its resistance to penetration by bullets fired from a range of

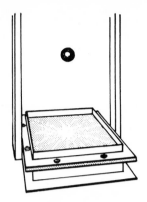

Figure 6.4 A dropping ball test for bandit-resistant glazing sheet materials.

guns. These are shown in table 6.1. Classes for glazing in the G category shown in the table are progressive, i.e. a panel complying with the requirements of class G2 also complies with the requirements of classes G1 and G0, but the S category is separate. There are two parts to BS 5051. Part 1 (1973) covers glazing for interior use, Part 2 (1979) covers glazing for exterior use. Both kinds of glazing must be exposed to external weathering before testing: interior grade for six months and exterior grade for twelve months. Test pieces of the exterior grade must also be conditioned, three at low temperature $(-20 \pm 2°C)$ and three at high temperature $(40 \pm 5°C)$ with the pattern of strikes completed within ten minutes from the time of removal from the conditioning temperature. Actual weapons are not used for the tests; special ballistic proof testing equipment is used and the mass and velocity of the bullets carefully controlled.

Test pieces are deemed to withstand the specified level of attack if the projectiles do not pass through them and if flying fragments from the projectiles or the test piece do not penetrate a witness card 450 mm from the rear surface of the test piece. Glazing panels supplied as conforming with BS 5051 must be indelibly marked by the manufacturer.

6.4 Choice of glazing sheet materials

Requirements other than safety and security that have to be taken into consideration when choosing a glazing sheet material include the following:

- Optical quality
- Stability
- Durability
- Maintenance
- Fire resistance
- Cost

Table 6.1 Bullet-resistant glazing test*

Class	G0	G1	G2	G3	S
Weapon type and calibre	Hand gun 9 mm military parabellum	Hand gun 357 magnum	Hand gun 44 magnum	Rifle 7.62 mm	Shotgun 12 bore magnum (full choke)
Ammunition	9 mm Mk 2Z standard	Soft point, flat nose, 10.2 g (158 grain) bullet	Soft point, flat nose 15.6 g (240 grain) bullet	NATO standard 7.62 mm ball	76 mm (3 in.) magnum cartridge, high speed black no. 6 shot. Mass of load 39 g (1.38 oz)
Test range	3 m	3 m	3 m	10 m	3 m
Number of strikes	3	3	3	3	2
Pattern of strikes (within practical limits)	Centres 100 mm apart forming an equilateral triangle within a square of 200 mm located centrally on the test piece			Centres coincidental and central on the test piece	

* Adapted from BS 5051

The requirements of sound insulation and thermal insulation seldom enter the criteria of selection because there is little difference in these properties of alternative materials in the thicknesses normally used.

Glass can be manufactured in forms that meet the requirements listed above and fulfil the functions given in section 6.2, as necessitated by the location of the glazed area.

Wired glass
Wired glass is cast glass with a wire mesh embedded in it. It is obtainable in the rough cast, translucent form and the polished, transparent form. The mesh does not add to the impact strength of the glass but it holds the glass together and breaks itself only under a severe blow. It therefore provides some safety and security in that it resists penetration. In a fire, the mesh retains the glass in position if it cracks from the heat of the fire. Cast wired glass 6 mm thick is not a safety glazing sheet material, however when polished it improves to Class C, possibly because of the removal of fine surface cracks from which fracture starts. As a security material it is resistant to the less determined burglar.

Toughened glass
Toughened or tempered glass is made from annealed glass by subjecting it to heating and rapid cooling, which induces high compressive stress in a thin layer at the surface of the glass with a compensating tensile stress in the interior. The glass can easily sustain this tensile stress. A stress equal and opposite to the 'toughening' stress can then be caused at the surface without putting the material in tension. Because glass is much stronger in compression than in tension, its strength is considerably increased by the toughening process, strength increases in bending of four or five times being achieved. When the glass breaks, the release of elastic energy stored in the glass by the toughening process causes multiple crack-branching and the glass breaks into comparatively small cuboid pieces. Once toughened, the glass cannot be cut or worked, therefore the size of the panel must be predetermined. Except for wired glass, all types of flat glass — sheet, float, polished plate, patterned and tinted — can be toughened.

Laminated glass
Laminated glass consists of two or more sheets of glass with layers of tough, transparent plastic, e.g. polyvinyl butyral, interleaved and bonded to the glass. On breaking, the glass particles remain firmly attached to the plastic which can absorb a considerable amount of energy with about 400% elongation before failure. Any type of glass can be laminated provided it has a smooth side in contact with the interlayer, but annealed glass is generally used. Three-ply laminated glass can be cut to size after manufacture without special equipment. Multi-laminated glass of five plies and over must be cut with a diamond-tipped saw. For safety purposes

three-ply laminated glass is used, usually with a 0.38 mm (0.015 in.) thick interlayer. Where there is a high risk of danger and it is important that the pane should strongly resist penetration, the tougher lamination with a 0.76 mm (0.030 in.) interlayer should be used.

Though laminated glass is safer than unprocessed annealed glass, it is as easily cracked so will be subject to the same possibility of needing replacement. Figure 6.5 shows thicknesses of glasses of different types having about the same resistance to impact breakage. The actual size sections in the figure will assist readers unfamiliar with metric dimensions to visualise thicknesses normally involved in glazing. In spite of 4 mm toughened glass being equal in strength to other thicker glasses, if toughened glass were to be substituted for them it would generally be used in a thickness of 6 mm to obtain the benefit of its greater strength and perhaps avoid the cost of replacement.

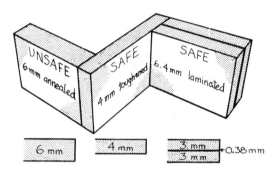

Figure 6.5 Glasses of roughly the same resistance to impact breakage.

Laminated glass can have wires embedded in the interlayer for decorative effect and to make the pane of glass more visible. One of the sheets of glass in a laminate can be wired glass, to make a fire-resistant safety glass. Wires that will form part of an intruder alarm circuit can be embedded in the interlayer. Alternatively, an almost transparent electroconductive metal oxide film can be deposited on the inner face of a glass ply. Cracking of the glass alters the resistance of the film to the current passing through it, causing an alarm to be triggered off.

Resin laminated glass is produced by a process much newer and simpler than that used for ordinary laminated glass. Two sheets of glass are held apart by edge strips while the space between them is filled with polyester resin which bonds to the glass on hardening. A class B safety glass can be produced by this process.

Anti-bandit laminated glass that will withstand criminal or vandal attack with iron bars, sledge-hammers, pickaxe handles and bricks, has a thicker interlayer than that used for safety glazing. In a typical three-ply construction, a 1.52 mm thick interlayer is used with 3, 4, or 5 mm glass. The interlayer is resistant to penetration of the weapons, though the glass breaks easily. Where people are likely to crack the glass through accidental contact, as in a shopfront or balustrade, the thicker glass plies are used. For greater protection against attack, a five-ply construction of 3 mm glasses and 1.14 mm interlayers is available.

A plastics and glass composite is also available for anti-bandit glazing. Acrylic sheet bonded to glass with polyvinyl butyral provides high resistance to attack by iron bars, pickaxe handles and the like. Because it is on the 'safe' side the acrylic sheet protects against spalling, where glass splinters fly off with particular danger to the face and eyes of persons behind the glazing panel.

Where bullet resistance is required, laminated glass made up as shown in fig. 6.6 may be used. If hit by a bullet on the 'attack' side, successive layers of glass are broken but the glass becomes very finely granulated at the point of impact and absorbs the energy of the bullet. Glass and interlayers may be of varying number and thickness according to requirements. A thick interlayer on the 'safe' side absorbs the shock waves produced in the material by the impact. A very thin layer of glass 1—1.5 mm (0.04—0.06 in.) thick limits spalling because it tends to powder with the shock of impact.

With some bullet-resistant glazing a layer of polycarbonate (described below) bonded to the laminate on the safe side prevents spalling. The use of polycarbonate results in a thinner and lighter panel for the grade of protection required, but the polycarbonate lacks the brilliance and resistance to scratching of glass. A low-spall grade of bullet-resistant glazing has the polycarbonate layer protected from maintenance problems by a thin layer of glass. Another

Figure 6.6 A typical 30 mm (1³⁄₁₆ in.) thick bullet resistant glass.

solution is to form what is known as a duplex panel and have a separate polycarbonate screen to intercept flying splinters. This can be replaced if necessary, comparatively cheaply, though it is a cumbersome solution to the problem and probably initially more expensive than other methods.

The thickness of bullet-resistant laminated glass varies from 30 mm (1 in.) for G0 standard to 50 mm (2 in.) for G3 standard (see table 8.2, section 8.11). Good protection against blast is given by laminated glass as shown in table 6.2. The test results in the table were obtained by the Department of the Environment, when investigating the effects of incident blast waves of high explosives. Since terrorist bombs rarely have even 30% of the efficiency of high explosives, these were very stringent tests.

To withstand blast equal to that of the 22.7 kg (50 lb) of high explosive used in the tests at 6.1 m (20 ft) from the glass, no single pane of 11.5 mm (0.45 in.) thick laminated glass should exceed 3 m^2 (32 ft^2) in area. The minimum edge cover should be 35 mm (1$\frac{3}{8}$ in.), an adhesive glazing compound should be used and the strength of the framing and its fixing should be engineering-designed for each individual system and set of circumstances.

For blast resistance of panes up to 1 m^2 (11 ft^2) a minimum thickness of laminated glass of 7.5 mm (0.30 in.) is recommended,

Table 6.2 Blast resistance performance of laminated glass (Impactex)

Glass		Explosive charge		Distance		Results
mm	(in)	kg	(lb)	m	(ft)	
4.0 (0.16) sheet*		0.7	(1.5)	1.22	(4)	Debris travelled up to 15.25 m (50 ft)
6.8	(0.27)	0.7	(1.5)	1.22	(4)	Glass cracked but stayed in frame; very small amount of fly-off
7.5	(0.30)	0.7	(1.5)	less than 1.22	(4)	As above; some fly-off
7.5 (0.30) with applied film**		0.7	(1.5)	less than 1.22	(4)	Extensive cracking of glass; interlayer did not rupture; film split but no fly-off
11.5	(0.45)	22.7	(50)	6.10	(20)	Glass retained in frame; some fly-off

*performance of monolithic glass given for comparison only
**0.05 mm (0.002 in.) polyester film bonded to interior face of glass with water-activated pressure-sensitive adhesive (see section 6.7)

and for up to $2 \, m^2$ ($22 \, ft^2$) a minimum thickness of $9.5 \, mm$ (0.37 in.).

Plastics glazing sheet materials

Though much tougher than glass, plastics used for glazing are on the whole optically less satisfactory, they are not so stable, not nearly so durable, they need special care in maintenance, they are combustible, and they are more expensive. Whereas glasses do not differ much in chemical composition, plastics can be formulated in a variety of ways. To the basic plastics mixture are added stabilisers, fire retardants and other ingredients. The resultant material will possess some qualities not present in a similar material but it is likely to be lacking in other qualities, for example fire resistance is obtained at the expense of weather resistance, impact strength at the expense of rigidity.

The optical qualities of a glazing sheet material depend not only on light transmission but also on flatness and rigidity. Plastics are flexible under stresses that would shatter annealed glass, unfortunately this flexibility causes distortion of vision. The coefficient of linear expansion of plastics used for glazing is about nine times that of glass; if the material is restrained by fixings expansion may lead to distortion. Different temperatures on either side of a glazing panel can cause bowing, as can also a humidity differential. Additionally, expansion and contraction can lead to broken seals that cause leaks.

The most serious degradation of plastics glazing sheet material is brought about by sunlight. Plastics are organic materials, when exposed to the weather they deteriorate like other organic materials. Ultra violet (UV) radiation initiates chemical reactions which are accelerated by heat and moisture. Yellowing, surface dulling and loss of impact strength are the result. The life expectancy can be improved by including UV stabilisers in the formulation, nonetheless particular care is needed in the selection of plastics for glazing in locations where temperature and humidity are high.

Plastics do not have the same resistance to scratching as glass. Near door handles and other places where users are likely to come into contact with a glazing panel the material can be badly marred. Abrasive materials must be avoided in cleaning. Even dust abrasion can lead to a serious loss of optical quality. Plastics suffer from dustiness because they carry a strong static charge. Scratches on the surface and chips on the edges will weaken the material. If it is fractured by an impact, cracks will tend to propagate along the scratches. Should a plastics glazing sheet material be fixed in contact with other plastics then plasticisers (substances added during manufacture to modify flow properties) may leach into the sheet material and weaken its edges. Organic solvents in paints and paint strippers may also damage the material.

Generally the choice of a plastics glazing sheet material for flat glazing will lie between the following:

- Polycarbonate
- Acrylic (polymethyl methacrylate)
- PVC (polyvinyl chloride)
- Polystyrene (often called styrene)

The safety and security properties of these materials are given in table 6.3.

Wired plastics are available for vertical glazing or roof lighting. A robust glazing sheet material used mainly for industrial buildings is a glass reinforced plastic (GRP) with polyester as the base material and an embedded mesh of expanded metal. Cellulose acetate butyrate (CAB) is another material for industrial buildings.

Corrugated sheets of acrylic, PVC and GRP have been used for roof lighting for some time. Optical requirements are less demanding and satisfactory service for good quality materials of forty years for acrylic and more than ten years for PVC and CRP can be expected before light transmission is seriously reduced.

There are some problems in determining the impact safety classification of plastics for vertical glazing. The loss in strength brought about by UV radiation lowers the resistance to penetration below what would be obtained in tests with new material. Also there is the difficulty of deciding whether or not a plastics material has broken safely. When assessing test results opinions may differ as to the danger of a pointed piece of broken material. The question of whether a definition of how pointed a piece must be to be dangerous can be formulated and incorporated in BS 6206 is under consideration.

6.5 Sheet materials for clear, vertical glazing compared

A comparison of sheet materials for clear, vertical glazing in ordinary situations is provided in table 6.3. Annealed glass is included because it is the standard material by which all other glazing sheet materials are judged.

Toughened glass in all thicknesses will break safely, but when it is broken its value as a protective barrier against falls or unauthorised entry is lost. Laminated glass continues to give containment after it is broken, but it breaks more easily than toughened glass. If laminated glass is made from toughened glass, containment and resistance to breaking will be provided but this is likely to be an expensive solution. When both no-break and containment are needed, plastics might meet the requirements better. Glass is a heavy material, plastics have about half the relative density and thinner sheets can be used, this might be another reason for choosing a plastics material.

Unfortunately, no plastics material is the equal of glass for retention of original appearance, resistance to weather and toleration of repeated cleaning. Acrylic and polycarbonate are the most widely used plastics with polycarbonate most favoured. Polycarbonate is by far the toughest, it will meet the requirements of BS 5544 for

Table 6.3 Sheet materials for ordinary, clear, vertical glazing

Properties — Material	Annealed float glass	Wired glass Cast / Polished	Toughened glass	Laminated glass	Acrylic Cast	Acrylic Extruded	Polycarbonate	PVC	Polystyrene
Optical quality	The standard	Mesh visible	Slight deviations from flatness Strain pattern seen in strong sunlight	As glass used	Almost as glass	Marginally less than cast — Not same surface brilliance as glass in service →	Marginally less than cast acrylic	Slight bluish tint	Distortion likely
Impact safety Classification (BS 6206)*	Unsafe	Unsafe Class C	Class A	Class C to Class A	Class C		Class A	Class C	Unsafe
Comparative resistance to impact breakage (very approximate)	1	1	8	As glass used	6—65		250	30	1—2
Provides containment if accidentally broken	No	Yes, unless heavy blow	No	Yes	No, but high-impact quality unlikely to break		Unbreakable	Breakage unlikely	No
Security resistance and resistance to breakage by vandals	Unclassified	Casual burglar	Casual burglar	Casual to determined burglar	Casual to determined burglar Vandal		Bandit Determined vandal	Casual to determined burglar Determined vandal	Unclassified
Abrasion resistance	← Not normally marred →			→	Fairly high		Fair Mar-resistant quality fairly high	Fair	Fair
Resistance to discolouration	← Not affected →			→	High		Fair	Fair	Fair
Resistance to crazing	← Not affected →			→	High	Fairly high	High	Very high	Fair
Resistance to dustiness	← Not affected →			→	Fair		Fair	Fair	Good
Resistance to solvents and paint strippers	← Not affected →			Interlayers damaged by oil	Fair to low	Low	Low	Low	Low
Weather resistance	← Not affected →			→	Fairly high 15—20 y	Fair 10 y	Fair	Fair	Interior use only
Thermal movement 60°C (108°F) range	← 0.5 mm per metre length 0.03 in. per 5 ft length →				← 4 mm per metre length 0.25 in. per 5 ft length →				↑
Fire resistance Class — BS 476 Part 7		Incombustible 1 hour fire resistance BS 476 Part 1		Interlayers are combustible	Class 3 self-extinguishing grades available		Combustible Class 1—2 self-extinguishing	Class 1 self-extinguishing	Class 3

*Probable test results for normal glazing thickness of plastics

resistance to manual attack by bandits, and when laminated to a thickness of 33 mm it provides bullet resistance to G2S (see table 6.1). A mar-resistant quality, produced by coating the sheet with a thin film of hard plastic, gives improved resistance to scratches and abrasion.

Cast acrylic will give the same clarity (initially, at least) as glass and it will be at least as strong as toughened glass in the same situation — one brand is said to have a breakage resistance seventeen times greater than sheet glass of the same thickness for the ordinary quality of acrylic and sixty-five times greater for a high-impact resistance quality.

When abrasion and scratching are not a problem the lifetime of cast acrylic should be ten or fifteen, possibly even twenty years. Extruded acrylic is an alternative material, it is marginally inferior to cast acrylic in strength and optical qualities, more susceptible to crazing and attack by solvents and paint thinners, but cheaper. Polystyrene sheet is cheaper still, it lies between annealed and toughened glass in resistance to impact, has less resistance than acrylic to scratching and is more susceptible to crazing.

With PVC a greater loss of definition compared with acrylic and polystyrene probably rules out its use where any obstruction to vision is displeasing. The life expectancies of polystyrene and of PVC are less than that of acrylic, particularly when exposed to the weather.

Plastics glazing sheet materials are combustible materials, thus when selecting one for a specific purpose consideration must be given to the way it will behave in a fire. The surface spread of flame test (BS 476 Part 7) puts materials into one of four classes. Class 1 materials make the least contribution to flame spread, they are described as having surfaces of 'very low' flame spread. The other classes are 2: 'low', 3: 'medium' and 4: 'rapid'. In addition to its combustibility, the thickness of a sheet material has a very important influence on the rate of flame spread. The temperature of the surface of the sheet has to be raised to the point at which it will ignite; the thicker the material the greater the amount of heat required. Generally, standard acrylic and polystyrene will be placed in class 3, which shows them to perform about the same as plywood and hardboard painted with oil base or polymer paints. Additives can give acrylic self-extinguishing properties, this means they will not continue to burn once the source of heat is removed, but this is at the expense of weather and humidity resistance. PVC and polycarbonate are self-extinguishing and very low spread of flame (class 1) materials. The combustible nature of plastics glazing sheet materials makes all of them vulnerable to one variety of vandal attack — they can be burned by cigarettes.

6.6 Thicknesses and glazing

Guidance on thicknesses of plastics glazing sheet materials to use is given in BS 6262 *Glazing for Buildings*. For general purpose internal glazing the recommendation is to use a minimum thickness of 3 mm for panes up to 1000 mm (3 ft 3 in.) on their longest side and 5 mm for panes with an edge greater than 1000 mm. Security glazing may have to be thicker, according to the material and the degree of protection deemed necessary. A manufacturer of polycarbonate gives as a general guide thicknesses as above for up to 1200 mm on the longest edge, 6 mm up to 1400 x 3000 mm (4 ft 7 in. x 9 ft 10 in.) and 8 mm for 2050 x 3050 mm (6 ft 9 in. x 10 ft).

In the fixing of glazing panels, the greater thermal movement of plastics compared with glass necessitates greater edge clearance and edge cover. Table 6.4 shows the recommended minimum edge cover for glass and plastics for safety and security applications. The edge clearance necessary will depend on the dimensions of the pane, but will need to be more for plastics than for glass. Where containment is one of the functions required of a thin plastics panel, deeper edge cover is advisable to prevent failure by displacement occasioned by the flexibility of the material. Blast-resistant glass in windows and shop fronts requires up to 35 mm edge cover. For general purpose, external glazing the sheet material will need to be stout enough to meet wind loading requirements as well as safety and security needs. Information on adequate thicknesses for wind loading purposes is given in BS 6262 and by manufacturers.

There are locations where a lightweight plastics glazed panel that could be easily displaced would be very acceptable. They occur where precautions are taken to vent a possible explosion. To reduce the magnitude of a gas explosion, the panel should fail or be bodily ejected at pressures not much in excess of 5 kN m^{-2} (104 lbf ft^{-2}). Among buildings where such panels might be installed are industrial buildings where gas is used as a fuel and, possibly, the kitchen of a flat. The risk of an accidental gas explosion producing forces significantly greater than those normally taken into account in building design occurring in any one flat in a block of 110 flats in a life of sixty years has been estimated at 1 in 40. (Taylor, N. and Alexander, S.J. (1974) 'Structural damage in

Table 6.4 Minimum edge cover for glazing sheet materials

Function	Minimum edge cover*			
	Glass		Plastics	
	mm	(in.)	mm	(in.)
Safety (internal)	8	($\frac{5}{16}$)	10	($\frac{3}{8}$)
Bandit resistance	12	($\frac{1}{2}$)	25	(1)
Bullet resistance	20	($\frac{3}{4}$)	25	(1)

buildings caused by gaseous explosions and other accidental loadings.' *B.R.E. Current Paper CP 45/74.*)

6.7 Plastics film for safety and security glazing

Glass may be covered with transparent or tinted polyester plastics film. The purpose is to hold broken pieces of glass together if the glass fractures. As a safety measure the film removes some of the danger from annealed glass, as a security measure it can prevent missiles or explosives thrown at a pane from penetrating. It will also reduce the amount of flying glass in an explosion. For anti-shatter purposes the film used is 0.5 mm (0.002 in.) thick, for anti-missile purposes 0.1 mm (0.004 in.) thick, or for extra protection 0.175 mm (0.007 in.). It can be applied in situ or before delivery.

Though developed for fixing to existing glass, the film has applications for new work. When a special glass is not available in toughened or laminated form, the film can be used to raise its safety and security value; it can, for instance, be used on fire-resistant glass. The film must be treated with the same care as plastics glazing sheet material, to avoid deterioration of its appearance by abrasion and scratches.

Chapter 7

The Perimeter

SAFETY

7.1 Boundary walls and fencing

Walls and fencing fail to fulfil the safety functions of perimeter guarding if they can be climbed or penetrated by children. Whether the dangers — busy roads, steep banks, watercourses — are inside or outside the grounds of a building is immaterial to the function. Walls of perforated blocks and fencing with horizontal rails less than about 750 mm (2 ft 6 in.) apart providing climbing holds will not be child-resistant.

Spiked iron pales should not be used unless the spikes are at least 1.8 m (6 ft) above the ground. If they are lower, children may try to scale the palisade from boxes or seats. When a small girl climbed on some boxes that had been stacked against a spiked palisade in her school playground she slipped and the sharp end of one of the pales penetrated her skull to a depth of several inches behind her ear. Firemen cut off the pale and the girl was taken to hospital with it protruding from her head. She was marvellously saved by a surgeon's skill, but other children have died when arteries in their legs have been cut by spiked pales.

Upright fence members must be close enough to prevent children squeezing through. This is particularly important if a child will be in immediate danger if he penetrates the barrier. A fence by, say, a canal is a doubtful safety feature if a child can get through it but an adult cannot, and it is too high to scale. The requirement that a sphere of 100 mm (4 in.) will not pass through the barrier, as specified in the building regulations for balustrades, should be observed. Also it should be noted that the general requirement for a balustrade guarding a ramp, landing or floor is 1100 mm (3 ft 7½ in.) height and that BS 1722 (1972) *Chain Link Fences*, Part 1 gives 1200 mm (4 ft) as the height for fencing children's playgrounds. When fencing sloping ground there is a need for special care, as shown in fig. 7.1.

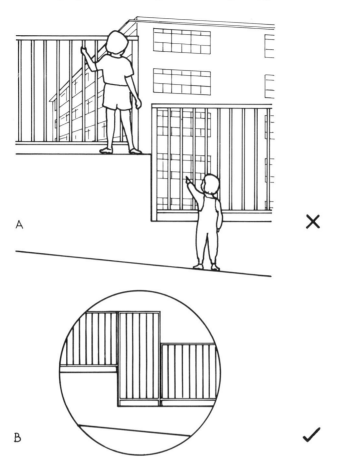

Figure 7.1 (A) Stepped fencing above a retaining wall fails to protect children from the risk of falling a considerable distance; (B) extending the upper level for 250 mm (2 ft 6 in.) past the step gives the protection necessary.

SECURITY

7.2 See-through or solid guarding

The question of whether perimeter guarding should be of the type that can be seen through or whether it should be solid depends on the situation and on what other security precautions are to be taken. Fencing that can be seen through allows a criminal to be observed before he can get inside and gives him no cover from outside observation once he has got in. If security guards or other people are on watch the criminal must evade detection by them even before he breaks in. If he gets through the perimeter he may be seen by the police or public when he attempts to get into the building through a door or window.

A solid and substantial high wall with spikes, barbed wire or Expamet anti-vandal scaling barrier (fig. 7.2) on the top may have the strength to resist attack (excluding explosives or ramming with a heavy vehicle) and be a formidable obstacle to scaling. But it will isolate the building and its surroundings from what is happening outside and, for what it is worth, provide the criminal with cover from outside observation. When the choice of see-through or solid perimeter guarding is considered, cost is usually the factor that rules out solid walls; this is not usually a matter of concern because the weight of the argument is likely to be in favour of see-through guarding.

Walls or close-boarded fencing may be chosen where there are requirements other than security to consider, such as privacy, reduction of noise level, formation of a windbreak. Solid guarding 1.6 m (5 ft 3 in.) high will give basic security and, if built taller with protection against scaling on the top and incorporating appropriate features from those described below for chain link and palisade fencing, maximum security can be achieved. Section 10.5 should also be consulted for security aspects of surface finishes and wall copings.

7.3 Fences

Fences may be constructed in many different ways; BS 1722 covers thirteen types, the most common that give at least basic security are chain link, close-boarded and wooden palisade. These are all capable of carrying barbed wire on top to give additional protection if required. Steel palisade fences used to provide strong to maximum security are also in common use.

Although basic security can be achieved with fences 1.6 m (5 ft 3 in.) high, 1.8 m (6 ft.) is to be preferred. I have seen a young man run straight at a 1.6 m fence constructed of steel angle with chain link infill and go over it without pausing. As he leaped at the fence he got a three-point grip on it, two hands on the top rail and the sole of one foot on the chain link, he brought his other leg up and vaulted over, holding on to the rail. Presumably he could

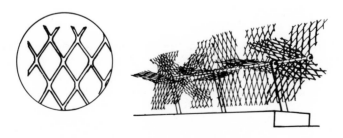

Figure 7.2 This scaling barrier revolves to prevent a stable hand hold, additionally the sharp ends of the mesh readily pierce gloves (the Expanded Metal Company Limited).

do the same with any other type of fence that was rigid enough and provided a suitable grip at the top. Close-boarded fences 1.6 m to 1.8 m high are not a difficult obstacle to an agile person if there is some object or another person to form a step or give a lift up. However the opportunist and petty thief will think twice before taking the risk of being trapped on the inside, even if he realises that the rails on the inside make the return easier. Wooden palisade fences give more protection if the pales are pointed.

Chain link fence

This is the most common type of fence used for security protection. BS 1722 Part 1 covers fences giving basic to strong security, BS 1722, Part 10 (1972) *Anti-Intruder Chain Link Fences*, covers the more secure type.

For maximum security a fence needs to be approximately 2.4 m (8 ft) high to the top of the chain link and 2.9 m (9 ft 6 in.) to the top of the barbed wire above the chain link. Ways that the fence may be attacked and precautions to take (figs. 7.3 and 7.4) are as follows:

- Climbing with toes in mesh — limit mesh size to 50 mm.
- Burrowing and lifting bottom of chain link — bury 300 mm (1 ft) of chain link in the ground, or thread hairpin staples through the bottom row of mesh and cast them into a concrete sill 125 mm (5 in.) wide by 225 mm (9 in.) deep, cast in situ for the full length between posts with the top of the sill approximately at ground level.

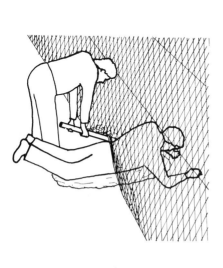

Figure 7.3 Some ways of circumventing a chain link fence.

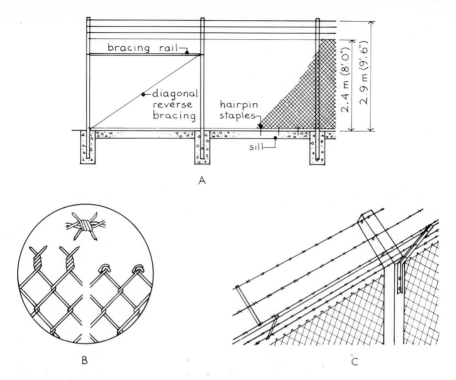

Figure 7.4 Measures to prevent the circumvention of chain link fencing. (A) General construction; (B) barbed and knuckle tops, and barbed wire; (C) posts with cranks both sides and spacing bars on barbed wire.

- Bunching the lines of barbed wire together and getting over between the barbed wire and the top of the chain link — fit spacing bars to barbed wire and use chain link with barbed top.
- Dismantling — burr bolt ends over nuts.
- Climbing straining posts — use a bracing rail and diagonal reverse bracing instead of a strut to strengthen straining posts; make changes in direction of fence not less than 130°.

These features should prevent an intruder breaking in without attracting patrolling guards or leaving obvious traces. If there is a danger of attempts being made to crash through with a vehicle, then either a trench or a stout kerb about 400 mm (1 ft 4 in.) high should be positioned where it will prevent a vehicle getting close enough to cause damage.

Steel palisade fences
These are strong fences obtainable from 1.2 m (4 ft) to 3.6 m (11 ft 10 in.) high (fig. 7.5). Corrugated or angle pales are used for fences up to 2.1 m (7 ft) in height, corrugated pales only for maximum security fences which should be at least 2.4 m (8 ft)

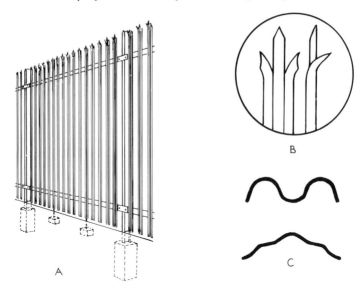

Figure 7.5 Steel palisade fencing. (A) A bay, showing R.S.J. section posts and intermediate supports for bottom rail; (B) splayed head; (C) typical sections.

high. I-section beams are used for posts, angles for rails. Galvanised and plastics-coated finishes are available. The fencing is designed for use without struts. Fixing bolts are burred over nuts to prevent dismantling, alternatively manufacturers use their own special vandal-resistant fasteners. BS 1722, Part 12 (1979) *Steel Palisade Fences* requires intermediate supports under the bottom rail but some fences are made with rails strong enough not to need this. The British Standard also requires a sill 125 mm (5 in.) wide by 225 m (9 in.) deep to be provided under the line of pales, the top of the sill to be approximately at ground level and not more than 50 mm (2 in.) below the bottom of the pales.

7.4 Gates for fences

Gates in a fence should be designed so that as far as possible they provide security comparable with that of the fence. In a chain link fence, gates are made of circular or rectangular hollow steel sections. For greater rigidity, 50 x 50 mm steel mesh welded at intersections is used instead of chain link infill. On double gates the closing leaf should incorporate a stop that prevents the drop bolt on the other leaf being lifted when the gates are closed. With wide gates the bolt should also support the closed leaves and be locked into the ground to prevent the leaves being lifted. A stout locking bar and good quality padlock should be fitted, otherwise thieves might drive up to the gates, attach a chain from their vehicle and pull off the padlock. The hinges must also be resistant to this sort of treatment. Barbed wire above the gate cannot usually be cranked because of fouling the cranks on the fence on opening or closing.

Gates for steel palisade fences are made with rectangular hollow steel sections welded together. The construction is not braced but corner strengthening pieces are used if necessary. Panels of rails and pales are then secured to the framing.

7.5 Special gates

Entrances will generally demand gates of greater architectural merit than those previously described. Often extremely wide gates are required. These are framed in trapezoidal section steel with, usually, bar infill. The gates may be hinged or sliding. If hinged they will require a large area for opening and closing and the road must have flanking walls or other boundaries for the gates to open against. The swing of the gate must be marked on the road surface to indicate that vehicles must keep clear. Hinged gates can be operated by electrical power, but the greater mechanical advantage of manual operation obtained by applying force as far as possible from the hinge means that heavy gates can easily be moved by hand.

Sliding gates are generally better than hinged gates for wide openings. They need less space and they are suited to powered operation. If the gate is very long, or very high, or perhaps very heavy because of armour plate infill it will need to run in a track that carries its weight. Otherwise a free-carrying gate can be used (fig. 7.6). This type has the advantage of being able to operate over surfaces at different levels and of being unaffected by snow and ice.

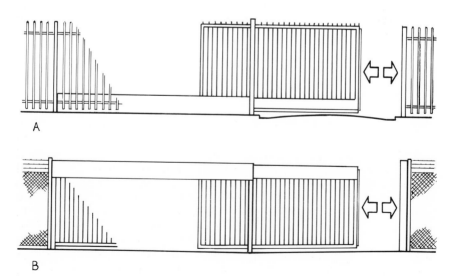

Figure 7.6 Free carrying gates. (A) Bottom supported; (B) top supported.

7.6 Entrance and exit control

Traffic barriers used at perimeter entrances and car parks are of light construction, intended only for the control of vehicles; however where there is a danger of vandalism, barriers of robust construction should be used. The lifting arm type is most common. Where space for installation is limited it is necessary to ensure that the arm can rise to the full vertical position or it may be damaged by high-sided vehicles or overhanging loads. The slewing arm type may be used instead where height is restricted. It is important that the arm slews away from the direction of the traffic, at least one incident has occurred where an arm slewing towards the traffic malfunctioned and a driver drove into it with serious results. An alternative for use with restricted height is the swan-necked type (fig. 7.7).

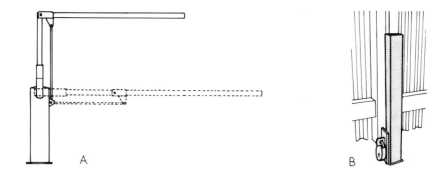

Figure 7.7 (A) Electro-hydraulic traffic barrier with swan-necked beam for restricted heights (Elkosta); (B) locking post (Sampson and Partners).

Rising road barriers (fig. 7.8) are also used to close roadways temporarily. These may not be in sight from within a vehicle, or they may not be noticed, so traffic lights are necessary to warn drivers when the barriers are raised. Unlike slewing and rising arm barriers, rising road barriers are resistant to vandalism and they can also stop a vehicle. The manufacturers of the large barrier shown in A, say that it will stop a 30 tonne truck dead, and it can be raised in two seconds. These large barriers can be used where there is a danger of kamikaze attacks, also where there is valuable merchandise that might be hi-jacked. Criminals may attempt to steal vehicles loaded with easily disposed-of goods, tobacco or whisky for example, and crash out through the gates. Rising road barriers facing inwards can be raised when the premises are closed to foil this form of theft. A cheaper device for the same purpose, effective against the lighter type of vehicle, is the lockable post shown at B, fig. 7.7.

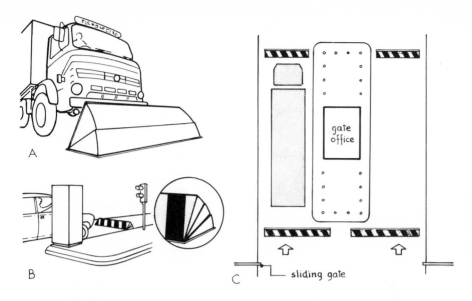

Figure 7.8 Rising road barriers. (A) Barrier to stop a 30 tonne truck (Frontier Gate); (B) barrier to control car parking (A.P.T.); (C) stop and search control.

Where maximum security requirements necessitate the inspection of vehicles entering or leaving the premises, barriers used in pairs, as shown at C, fig. 7.8, enable the passage of vehicles to be fully controlled, there can be no rushing of a lowered barrier by a following vehicle. Two entrances, as shown, keep vehicles off the road should a second one arrive before the first is dealt with.

The remote control of electrically operated gates and barriers is similar to that of doors, as described in section 13.11. One difference is the use of loops below the road surface to detect the presence of a vehicle or other large metal object, and initiate opening. (The metal object does not always have to be large, I know one that a workman makes respond to his shovel.) The loops are usually used at unsupervised exits where the purpose of the barrier is to bar the unauthorised entry of vehicles.

Methods of access control of pedestrians are described in section 13.11. Control may be at the perimeter for maximum security, and usually effected by a turnstile, card operated or supervised; more often access control is operated at building entrances, and for extra protection both perimeter and building access may be controlled.

Chapter 8

The Spatial Environment

8.1 Spatial considerations

The creation of a suitable spatial environment for safety and security evolves largely from consideration of the relationship between contiguous buildings and between parts of individual buildings. It comes also from special attention given to areas of buildings that are particularly vulnerable from the safety or security angle. Examples include kitchens for accidents, public toilets for vandalism and storerooms for theft. For safety, spatial relationships and special provisions must go beyond the avoidance of the more obvious dangers that lead directly to accidents, and provide good ergonomic design that will lessen the strains and irritations that lead indirectly to accidents in ways often difficult to define. For security, opportunities for vandalism and theft must be reduced by avoiding the creation of secluded places for wrong-doing; separating activities carried out in buildings (to facilitate control of and observation of users) and making special provision in vulnerable areas.

SAFETY

8.2 Safe routes

Footpaths should be sited well away from the walls of buildings to avoid the danger of pedestrians colliding with open windows (fig. 8.1) or up-and-over doors (fig. 12.16). They must also be kept clear of fuel supply inlets because of the danger of oil spillage. Handrailing or protective walling may have to be provided to footpaths near high-rise buildings because of the way these buildings increase wind speed around them. A single tall building causes the wind to speed up as it rushes around it attempting to maintain its place in the air mass; clusters of buildings have a funnelling effect leading to further increase in wind speed and causing gusts and eddies. Old and frail people are blown off their feet, sometimes with fatal results; they are particularly at risk if the paths are icy (fig. 8.2). Handrailing is also necessary when steps are required in a footpath because of changes in level. Gentle slopes are safer than steps, thus ramps are to be preferred.

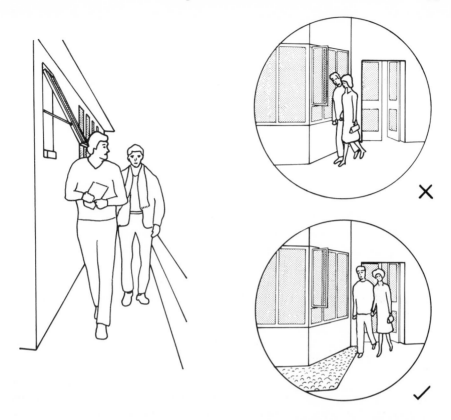

Figure 8.1 Footpaths must be kept at a distance from buildings, and pedestrian routes must be clearly defined, to avoid the hazard of open windows.

Vehicles should follow one-way routes in the surroundings of a building wherever this is practicable and should be segregated from pedestrians as much as possible. If pedestrians have to be in the same area as vehicles, as in car parks, main routes for people on foot should be clearly separated from vehicle routes in ways that do not introduce fresh hazards for pedestrians or vehicles (fig. 8.3). Raised paths for pedestrians are preferable, otherwise the paving should have a distinctly different surface from that of the vehicle area. Car park entrances and exits must be safe for pedestrians and vehicles; although there may be alternative exits for pedestrians, people tend to leave buildings by the way they enter them (fig. 8.4). If blind vehicle exits cannot be avoided, steps must be taken for the safety of people crossing the exit (fig. 8.5).

The possibility of an impact from a vehicle being severe enough to endanger occupants of the building is generally too remote to warrant precautions against it, but if a building is on a bend in a road used by juggernaut lorries at speed, or at the bottom of a steep hill, it may be advisable to form an earth bank or build an

Figure 8.2 Where the wind is strong the low rails used to keep people and vehicles off grass should not be used; old people are likely to fall over anything below knee height if blown into it by the wind.

extra strong boundary wall to absorb or deflect impact from vehicles out of control.

Routes within buildings should be as direct as possible. The dangers that might be met in having to pass through a laboratory or workshop to get to another room can be readily imagined. A kitchen is a workshop and like other workshops, should not be

Figure 8.3 The fall of the ground suggested this method of segregating vehicles and pedestrians: the hazard of the step was not foreseen.

Figure 8.4 People risk being struck by vehicles and traffic barriers at the vehicle entrance/exit of this car park because the pedestrian entrance/exit is sited where it is neither readily seen nor convenient to use.

designed as a throughway. The house shown in fig. 8.6 is a bad example but not atypical, the whole house is a throughway. Should the housewife be in the utility room when someone rings the front door bell, she will have to pass through every room on the ground floor to answer the door. Opportunities to trip over something or someone on the way will be plentiful. Five doorways to negotiate provide other possibilities for an accident.

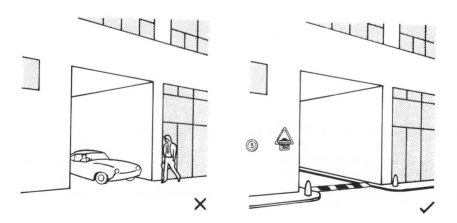

Figure 8.5 Danger at a blind exit, and precautions.

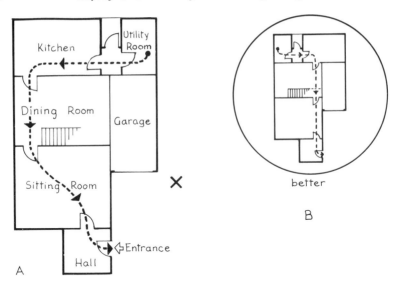

Figure 8.6 (A) In the absence of a communicating passage between rooms the whole of this house becomes a passageway; (B) with the same constraints on space this plan makes for shorter and less hazardous journeys.

8.3 Kitchens

For safety in kitchens the main requirements are sufficient space to accommodate activities and appliances, and a good ergonomic relationship between them (fig. 8.7A): requirements that are equally applicable to other workshops and to laboratories. Kitchen planning is based on the work centre, or activity zone, which is the place where a particular type of work is done. The place is named after the principal item of equipment in it, examples are: sink centre and cooker centre, alternatively it can be named after the type of work done, food preparation zone and cooking zone, for example.

The most important centre is the sink centre, it comprises sink, drainer, worktop, dishwasher and possibily storage for vegetables and refuse container. Next comes the surface cooking centre, surface cooking requires more attention than cooking in the oven so when a separate high level oven is selected that is best placed at the end of a line of worktops to avoid interrupting the run of work between surface cooking and other centres and also creating a hemmed-in feeling. On each side of the cooker or hob unit there must be a worktop at least 300 mm (1 ft) long to stop small children reaching up for saucepan handles (fig. 8.7B); island hob units must have a worktop along the back as well. Stirring tools and other items should be stored under the worktop or in cupboards or racks over them. Nothing should be stored above the hob — an item might fall onto utensils or the hob itself.

Studies carried out in well designed kitchens in the U.S.A. show that the most frequent shifts of attention ('trips' in kitchen research parlance) between centres occur with sink and surface cooking centres. Where a kitchen has hobs and oven combined in one cooker, trips between the sink and the centre where mixing is carried out ranked second (even for meals without baking) followed by trips between sink and refrigerator. Where hobs and oven were separate, the additional centre that resulted reversed the order of the second and third, bringing trips between sink and refrigerator into second place followed closely by those between sink and mix centres. The importance of the mix centre is declining as more ready-mixed and ready-cooked meals become available, but if a kitchen is planned for the more exacting requirements it is likely to prove convenient and safe for simpler needs. Hence a kitchen should be planned by locating centres in the following order of priority:

1. Sink
2. Cooker or hob
3. Mix
4. Other, e.g. refrigerator, china, eating
5. Separate oven and freezer

The strongest relationship should be between sink and hob centres, when that is achieved consider next obtaining a strong relationship between the sink, mix and refrigerator centres.

The principal piece of British official advice on kitchen planning is that the sequence of fitments should be **worktop/cooker/worktop/ sink/worktop (a drainer is a worktop), or the same in reverse order, unbroken by full-height fitments, doors or passage ways**. This

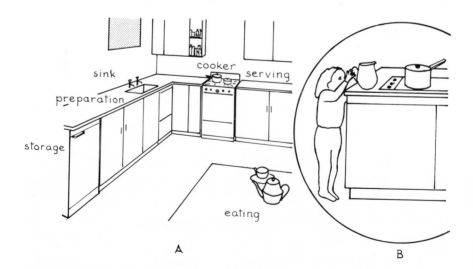

Figure 8.7 (A) Kitchen layout for efficiency and safety; (B) island hob units must have a worktop at the back as well as the sides to keep the more dangerous items out of reach of small children.

arrangement has been found to give good results in practice. The D.O.E. design bulletin on kitchens (No. 24, Part 2) emphasises the *work triangle* made up of the paths between sink, cooker and refrigerator or centralised food store. This triangle, it is suggested, should be between 3600 mm (12 ft) and 6600 mm (22 ft) to give sufficient working space and yet be reasonably compact. Centres may be arranged in a straight line or, providing there is no through-way to interrupt the work flow, in an L-plan or a U-plan.

Design based on the foregoing principles should avoid the undesirable location of cookers, such as near a doorway, where there is a danger that the door or the person opening it will come into contact with the person using the cooker (fig. 12.11). Another undesirable location is in the corner of a room, here access to some utensils on the hob will be only over the top of other items. Cookers should have worktops at hob level on both sides so that pots and pans can be slid from the hob directly on to the adjacent surface. Worktops on opposite walls must not be at different levels: when the housewife swings round from a lower surface to a higher one she may catch utensils on the edge of the higher surface.

Insufficient worktop space not only makes manoeuvring of ingredients and utensils difficult, but demands greater attention to finding space to put things, which can lead to accidents. Factors such as the number in the family and the amount and variety of cooking influence the length of worktop needed in a kitchen; in turn the length of worktop determines the amount of storage that can be provided in the counter underneath and on the wall above. And adequate storage to avoid clutter is important in accident prevention. Financial considerations are usually limiting factors, but the general consensus of what constitutes minimum to liberal worktop provision at centres is as follows:

Oven	300—450 mm (12—18 in.)
Hob or cooker	300—450 mm one side, 600 mm (24 in.) the other side
Mix	900—1050 mm (36—42 in.)
Sink	900 mm (36 in.) + drainer, or 900 + 750 mm (36 + 30 in.)
Refrigerator	300—450 mm (12—18 in.)

Centres may be combined to reduce the total length of worktop necessary, in which case the combined worktop need only be 300 mm (12 in.) longer than the longest single worktop that would otherwise have been provided.

Doors from a kitchen into another room (rather than into a hallway) and to outside take up space, yet they are desirable because they enable a mother to supervise her children at play while she works in the kitchen, and they give her quick and easy access to them when necessary. When the kitchen is combined with a dining or living area, a separating fitment that incorporates a small gate will assist in keeping toddlers away from the hazards of cooking procedures and the mechanical and electrical dangers associated with kitchen appliances.

8.4 Domestic bathrooms

There should be sufficient space in bathrooms and WC compartments to allow for the tending of children, old people and invalids. And grab bars should be provided for all ages of users: minimum grippable lengths given in the part of the American Society for Testing and Materials (A.S.T.M.) consumer safety specifications for bathrooms that covers grab bars and accessories (F446—78) are shown in fig. 8.8.

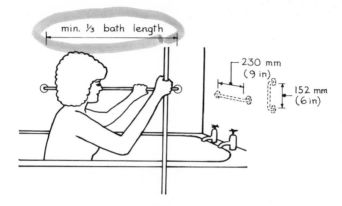

Figure 8.8 Grab bars are essential for bathroom safety; dimensions shown are minimum grippable lengths.

If an arrangement of bath, washbasin and other appliances that will allow a small child to climb from one to the other cannot be avoided (fig. 8.9) because of lack of space, then the child must not face additional danger at the top of the climb — an openable window or a medicine cabinet perhaps. (Safety organisations advocate lockable cupboards for medicines to make them inaccessible to children. Keys get lost or left in the lock, so locking devices operated by hand in ways intended to be too difficult or too mysterious for children are used. This has had little effect on the incidence of child poisoning — children who eat tablets thinking they are sweets usually find them by the bedside or in some other accessible place where they have been left ready for taking.)

Pull switches in bathrooms must be carefully located. Often they are placed where they are fouled by the door on opening; in a small bathroom where the full width is taken up by the doorway and the bath it is difficult to avoid this. Householders may then shorten the cord to stop it tangling with the door, putting it out of reach of children unless they put themselves at risk by climbing on the bath (fig. 8.9). A switch outside the bathroom is preferable.

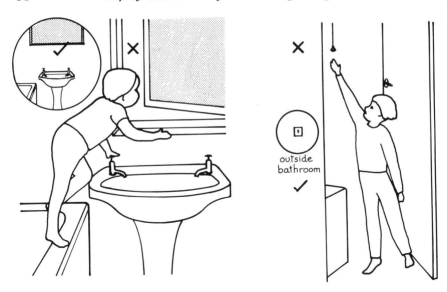

Figure 8.9 Avoid climbing routes, and the temptation to climb, in bathrooms.

Other safety requirements in a bathroom are safe heating and a door bolt that can be opened from outside in an emergency.

8.5 Pipe runs and service ducts

The need for airlocks and washing and changing facilities is not likely to be overlooked for laboratories where dangerous diseases are investigated, nor for hospital wards where patients with highly infectious ailments are nursed, but the run of service pipes and ducts from such places may not be given proper consideration. The case of smallpox, in which a service duct was thought to be involved, was mentioned previously; recently a British hospital was built with pipes from the pathology department running through the kitchens.

SECURITY

8.6 The site

Nooks and crannies are the workplace of the vandal, and choice entry points for the burglar if they contain doors and windows. Therefore, as far as practicable, buildings should be sited so that the whole of the exterior is in view from the road (fig. 8.10). If this is effected, the benefit must not be negated by subsequent planting of trees and shrubs that will provide cover for the criminal.

Except where roadways and footpaths must run up to entrances, they should be kept as far away from a building as possible to deny potential intruders an excuse for being in the vicinity of the building.

Footpaths should be within sight of occupants of buildings; tunnels, sudden turns round corners and other blind spots should be avoided to lessen the opportunity for attacks on pedestrians.

Car parks also should be well away from buildings and clearly visible. Employees' and customers' car parks are best separated from working and public areas of a building complex to restrict both internal and external theft (fig. 8.10). Car parks and other paved areas also attract ball players whose rough play verges on vandalism. Sometimes constraints of space and finance require a car park to be adjacent to a building housing items of value, in such situations thieves have broken in by repeatedly ramming the wall with a stolen lorry (fig. 8.11). A Land Rover with a steel beam in the back has been used in a similar way. A substantial dwarf wall or other barrier between the car park and the building is called for, otherwise the walls of buildings requiring strong security must be made stout enough to resist such attacks.

8.7 The buildings

Zoning, preferably with access control (see section 13.11), should be arranged so that the general public are restricted to parts intended for them and all but selected employees are kept out of computer rooms and other places where confidential files are kept. In addition, grouping together high risk areas will permit high grade physical security measures to be confined to a limited part of the building. The rest of the building need not bear the increased construction costs nor suffer the constraints on design that the higher grade of security demands.

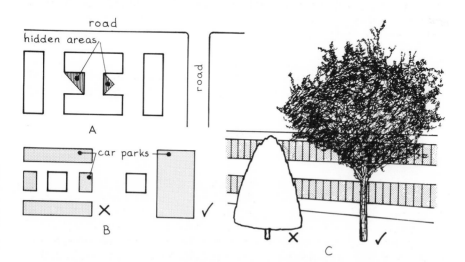

Figure 8.10 (A) Plan of building complex showing areas hidden from roads; (B) relationships of car parks to buildings; (C) choice of trees to avoid cover for the criminal.

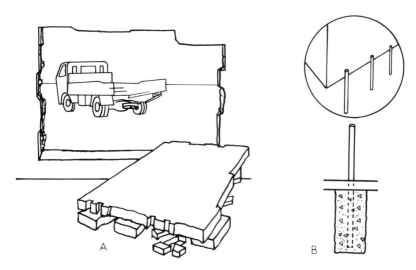

Figure 8.11 (A) Three impacts from a stolen lorry against the 275 mm (11 in.) cavity wall of a warehouse made a hole through which thieves took 500 000 cigarettes; (B) a preventive measure which also finds an application outside high-risk shop fronts.

A school is an example where targets can be brought together and given extra protection (fig. 8.12), though the same can often be achieved with commercial and other buildings. In a school the vulnerable places are the office and the head teacher's room, where cash and records are kept, and the storeroom where TVs, videos and other expensive items are kept. The toilets are also vulnerable to attack, though in this case when the building is occupied. They are included in the secure, administrative zone so that they can be kept under observation more easily, their supervision is also

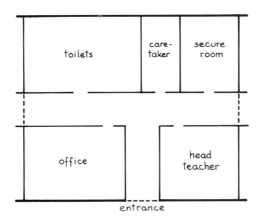

Figure 8.12 Grouping of vulnerable areas of a school.

enhanced by having the caretaker's room adjacent — the cleaning, unblocking and upkeep are his responsibility.

In addition to a higher grade of security construction in respect of the building envelope and in the doors and walls of the rooms in the security zone than elsewhere, the use of intruder alarms may also be confined to this zone, except where they are necessary to guard against nuisance and malicious attacks.

Because school and college buildings are used for public functions when the rest of the building is unused, zoning should ensure that the public can be confined to selected parts of the building. Conference suites in public buildings such as town halls should similarly be capable of isolation from the rest of the building.

Loading and unloading bays are vulnerable security points, hence bays should be sited where they are under observation from offices (fig. 8.13). Restrooms for drivers (who should not be allowed to load their own vehicles) should be located some distance from the loading bays. Security in the removal of scrap and rubbish requires precautions equal to those for the removal of goods.

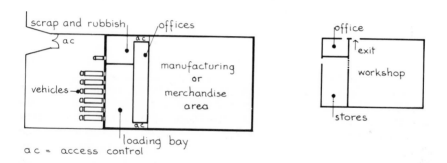

Figure 8.13 Planning that separates production from areas where pilfering is likely, and provides for observation from offices.

Production workers should not have access to storage areas (fig. 8.13). Cash offices, and secure rooms likely to be attacked when the building is open, should be placed well inside buildings, preferably on an upper floor, so that they cannot be reached by a straight and direct route, with the proviso that vehicles bringing or removing cash must be able to approach the building closely. This may require the use of removable road barriers such as the lockable post shown in fig. 7.7. The inadvertent design of climbing routes providing easy access to flat roofs and balconies (fig. 8.14) should be avoided, these provide opportunities for both vandals and burglars.

With the increase in occurrence of armed robberies in the U.K., this form of attack is likely to spread from the present targets in the jewellery trade to other shops and stores. In the U.S.A. almost

Figure 8.14 Climbing routes.

any type of store is exposed to armed robbery, usually for quite small amounts of money. Their experience shows that the gunmen seek out stores where the interior is not clearly visible from the street or car park, and where there are places to lurk unnoticed inside. A clear view into the store, and low-level display and shelf fitments — under 1.5 m (5 ft) — are a deterrent. Low-level fitments are now preferred in libraries so that those in charge have a better view of what is going on. A dais raised about 150 mm (6 in.) at the checkout point or a raised floor behind the sales counter assists staff in their observation and puts a potential robber at a psychological disadvantage.

8.8 Entrances and exits

In public buildings, illegal entry should be discouraged by having entrance doors under close observation. In buildings where there is no reception office the general office should be placed, and constructed, so that the entrance can be kept in view. Generally the day-to-day use of exit doors provided for escape in case of fire should be prohibited. They are a way out for persons engaged in theft (employees and outsiders) and a way in for intruders. Door furniture should be provided on the inside only, key holes should not penetrate to the outside.

From the outside, emergency exit doors should be in plain view so that they are less likely to be used for illegal entrance, at the same time the doors should not be located where the thief will have a clear and direct getaway. There will be difficulty in reconciling these two requirements; an alarm on the door and an enclosed route out that passes close to the security gate may be a possible solution in some instances. The doors should also be in a position where they can be observed from inside the building. Exit doors in stairway shafts pose difficulties in this respect; they should, wherever possible, be rendered visible by glazed doors giving entry to the shaft (fig. 8.15). Outside, emergency exit doors should preferably discharge on to a grassed area so that any person lurking around or leaving by the doorway is conspicuously out of place.

Most university, college and school buildings invite unauthorised entry by the number of entrances and exits they have. These establishments have been able to reduce the difficulties they have experienced because of the easy entry of intruders by permitting exit-only through most doorways. Much trouble would have been avoided had the buildings been designed with knowledge of security requirements. When exits separate from entrances are required to avoid congestion, automatic-opening doors that operate only from indoors may be installed.

8.9 Public toilets

Open access, seclusion and attractive targets with the possibility of aggravating damage by squirting water about or flooding the premises are enticements to vandalism in public toilets (fig. 8.16). Robust construction and the avoidance, as far as possible, of anything that can be swung on, stood on, sat on or pulled off are essential. Vandals should not be tempted to stand on washbasins. They will do this if washbasins are placed adjacent to cubicles or under high-level windows, and they play the fool by looking over the cubicle partitions and shouting at passers-by out of the windows. Wall and partition surfaces should be of vandal-resistant tiles or

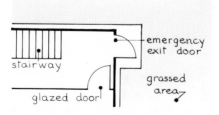

Figure 8.15 Security measures at emergency exit, permitting observation from inside, and drawing attention from outside the building, to unauthorised approach or exit.

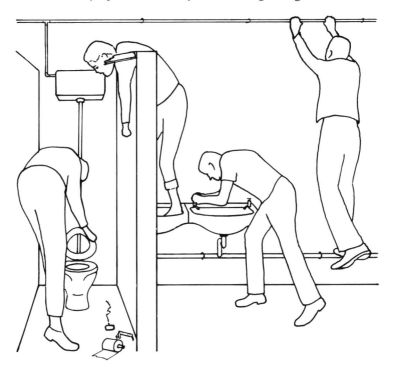

Figure 8.16 Some of the destructive antics of vandals in toilets.

steel, mirrors of stainless steel. The floor should drain to a gully and, to facilitate the unblocking of drains, WC pans should be provided with rodding eyes and a simple drain layout to manholes.

For a basic level of vandal resistance, in a school for example, fixtures must be well supported and pipework must be of a substantial size and fabricated where possible in steel pipes. Public or school toilets are not the place for plastic pipework or unscrewable bottle traps. For a higher level of vandal resistance, fixtures such as those shown in fig. 8.17 are required. These have concealed fixings and support brackets, button or foot operated valves, and ducted plumbing. Stall urinals are recommended by most authorities as pod urinals are sometimes wrenched from the wall, however stall urinals lend themselves to causing more trouble by blockage than the pod type — toilet rolls are dropped into the waste channels.

8.10 Secure rooms and strongrooms

Rooms where cash and records are kept or goods attractive to thieves are stored may be categorised, for convenience of description here, as storerooms, secure rooms or strongrooms, according to the grade of security deemed necessary. All these rooms should have finishings and fittings requiring the minimum of maintenance. This is to avoid the need for entry by people who would otherwise be excluded. For the same reason switchboards and alarm controls should not be located within the rooms; it may be advisable for these items to be in a protected place, but it must be a separate place.

The constructional features required are described in the chapters that follow. A storeroom will incorporate the essential items for defence against the opportunist and minor criminal. A secure room giving strong security needs a substantial door and frame, windows fitted with bars or grilles (or alternatively made windowless) partitions of brick or concrete, with a strength at least equal to that of 100 m (4 in.) reinforced brickwork, and external walls, floors and ceilings equally resistant to attack. Strongrooms for the protection of vital records and irreplaceable items need to be fire-resistant and capable of surviving intact even if the rest of the building burns down or collapses. The degree of burglar resistance required in addition to the fire resistance will depend on the nature of the items to be stored.

The reinforced concrete construction with mattress type expanded steel reinforcement as a barrier to attack (fig. 8.18), used for strongrooms built with the rest of a building, and the interlocking block construction built into existing buildings, may be replaced now where a small strongroom with maximum security is required by a prefabricated structure such as the Chubb Demountable Extendable Vault. This is made of 160 mm (6½ in.) thick concrete

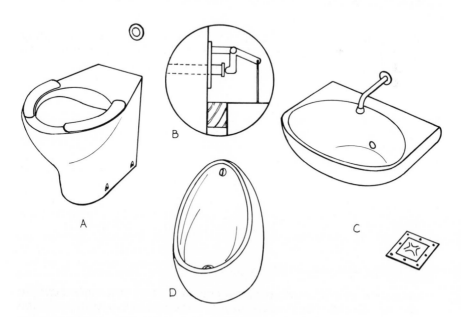

Figure 8.17 Anti-vandal sanitary fixtures (Armitage Shanks) with back to wall fixing, and concealed waste traps. (A) WC pan; (B) push button flush mechanism; (C) wash basin with foot operated valve; (D) urinal.

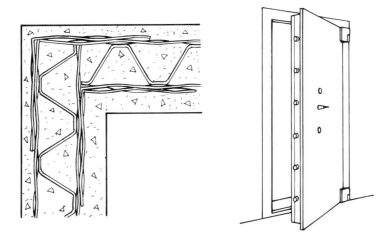

Figure 8.18 Corner of strongroom wall with mattress reinforcement and steel thief resisting door (Chubb).

of a special aggregate reinforced with steel and with polypropylene fibres. It is said to be resistant to all known forms of attack. The standard width is 2.48 m (8 ft 1½ in.) and it can be made up in various lengths. Besides the advantages of speed of erection, and less messy erection in an existing building, the prefabricated construction has the advantage of the tight quality control possible with factory manufacture.

Doors to rooms where security is important need to be stronger than the walls, floor and ceiling because if a thief can open a door he thus obtains better access than he would have through a small hole made somewhere else in the enclosure. Where a door stronger than those of ordinary construction described in section 12.8 is required, a door of special construction such as that shown in fig. 8.18 should be used having a face plate of steel up to 12 mm (½ in.) thick, and incorporating fire-resistant material. An enclosure of reinforced concrete would need to be at least 225 mm (9 in.) thick for comparability with this door. A strongroom with such construction and door will give adequate protection for important records and limited quantities of valuables. To withstand a determined attack by professional criminals skilled in the use of techniques involving the use of explosives such as 'shaped' charges, used in the demolition of buildings, diamond drills and thermic lances, the construction must either be more massive or special materials such as that used for the demountable vault described above must be used.

Rooms of above average security are required not only for the storage of records and valuables — dispensaries, weapon storage areas, and central alarm stations are among sensitive areas needing special protection. If purpose-built, these rooms may be constructed as just described, if up-graded after construction the rooms are often strengthened with steel sheets. Composite panels of steel and polycarbonate (fig. 8.19) tested by the U.S. Naval Civil Engineering Laboratory ((1982) 'Increase security in sensitive areas'. *Techdata Sheet 82–04*. Published in Port Hueneme) have greater resistance to penetration than solid steel of the same thickness, though they are only half the weight. The denial time (expressed as the time taken to cut a 250 x 250 mm (10 x 10 in.) hole using a reciprocating saw, an abrasive wheel saw or an oxyacetylene torch) compares favourably with the denial time of 200 mm (8 in.) reinforced concrete, which is 30–60 minutes.

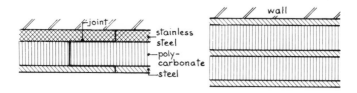

Figure 8.19 Composite intruder-resistant wall panels of 6.4 mm (¼ in.) stainless steel, 12.7 mm (½ in.) polycarbonate and 3.2 mm (⅛ in.) steel.

The panels are advantageous for installing in existing buildings. The components can be carried in separately and installed one layer at a time, after cutting and drilling has taken place elsewhere. Obviously the polycarbonate need not be of glazing quality, the sheets are bolted together with the steel sheets and bonded with polyurethane adhesive. Panels must be well fastened to each other and securely fixed in place using steel sections.

The five-layer panel will stop a 7.62 mm NATO round fired at a range of 25 yards (22.86 m) but the three-layer panel will not. For comparison, in the BS 5051 test for class G3 bullet resistance, the range for a 7.62 mm round is 10 mm (see table 6.1).

8.11 Secure offices

Protection for personnel is the principal difference in the security requirements of secure offices compared with secure rooms. When face-to-face contact with potentionally hostile people outside the office must take place, personnel need to be able to communicate without risk of being attacked. If cash and valuables are to be protected, personnel must not be put at risk of injury for not handing over property or allowing entry.

Communication and the transfer of cash can take place via small windows, this is suitable for intra-company transactions such as the payment of wages or the receipt of takings; for transactions with the general public a more welcoming approach is preferred. The cash office shown in fig. 8.20 has payments windows, these can be glazed according to the grade of protection considered

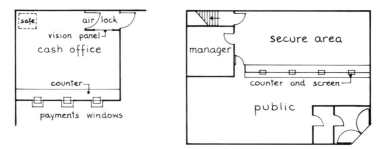

Figure 8.20 Planning for the security of personnel and cash in places where payments are made and received.

necessary and fitted with shutters closable from inside. To prevent robbers bursting in when the door is opened for receiving bulky items, or for other purposes, an airlock is formed. Electric locks or electric releases, in conjunction with a control unit, will prevent one door being opened before the other is closed (see section 13.10). Alternatively, bulky items such as security bags and wages boxes can be passed through a hatch with doors on each side. These doors, also, should interlock so that both doors cannot be open at the same time.

In offices where an ostensibly open contact with the public is required the secure area will be separated from the public area by counters and glazed screens. The construction of these will vary according to the grade of security they are required to provide, the principal features are shown in fig. 8.21. In a benefits office, laminated safety glass in the screen and a reasonably stout counter front should be sufficient; in a bank, bullet-resistant glazing and a bullet-resistant counter front and top might be considered necessary. The upper screening can probably be omitted where the lowest grade of security is sufficient; for strong security a grille resistant to forced entry can be used; for maximum security, precautions must be taken against a robber standing on the counter and aiming a gun at personnel in the secure area. Elsewhere, gaps through which a gun could be aimed must be avoided.

Tests for classifying the bullet resistance of glazing materials and the materials available are described in sections 6.3 and 6.4. Glazing to give high resistance against bullets is extremely heavy (table 8.1), possibly requiring support from a steel beam underneath the counter top.

Maximum security, or background noise as in booking halls of major rail and coach terminals may demand that communication from one side of the screen to the other be by microphone and loudspeakers. Without electronic aids, there is less distortion of speech when sound waves are reflected through gaps at the sides and top of the glazing than when they have to find a path through overlapping glazing panels.

There is a danger that if a gunman cannot intimidate staff behind bullet-resistant screens, he will grab a hostage and threaten to shoot him unless money is passed over. To counteract this form of attack the Fichet-Bacche 'Guichet' system of motorised armour-plated screens has been developed. Normally no screens are visible, allowing an open personal relationship with customers. When an attack is threatened the screen rises to full extension in 0.7 seconds, completely closing off the staff from the customers.

Steel and aluminium armour, glass reinforced plastics (GRP) and compressed hardwood plywood are used for bullet-resistant panels, partitions and doors. Thicknesses of these materials to resist the test firings of BS 5051 are given in table 8.1. A British Standard for opaque bullet resistant materials is under consideration; until this is available the standard for bullet-resistant glazing has to suffice for the comparison of the various opaque materials. The thicknesses given in the table are typical minimum thicknesses. The way in which the material is used will affect its performance — for instance a backing material may affect the way that the bullet-resistant material dissipates the energy of the bullet in bringing it to a stop.

The weight advantage for bullet resistance lies with steel armour followed closely by GRP until resistance against heavy attack is required, when aluminium armour comes second. There are of

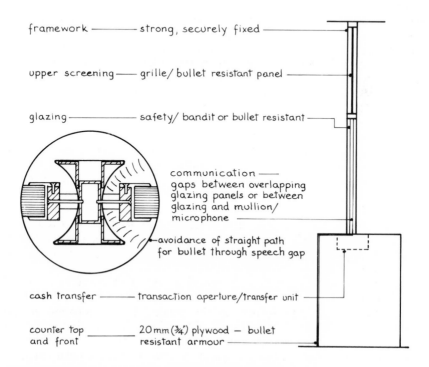

framework ——————— strong, securely fixed ——————

upper screening —— grille/ bullet resistant panel ——————

glazing —————————— safety/ bandit or bullet resistant ——————

communication ——
gaps between overlapping glazing panels or between glazing and mullion/ microphone ——————

—avoidance of straight path for bullet through speech gap

cash transfer ——————— transaction aperture/transfer unit ——

counter top
and front _____ 20 mm (¾') plywood — bullet resistant armour ——————

Figure 8.21 Requirements of security counters and screens.

Table 8.1 Typical thicknesses of materials for bullet resistance to test firings of BS 5051

Class	Reduced spall laminated glass (Impactex)	Glass and polycarbonate composite (Omni-armour)	High hardness steel armour (Compass B555)	Aluminium alloy armour (Alcan Plate)	GRP (Bristol Armour)	Compressed hardwood plywood (Delignit)
Thickness in mm (in.)						
GO	30 (1.2)	20 (0.8)	2 (0.08)*	7 (0.3)	7 (0.3)	30 (1.2)
G1	35 (1.4)	24 (1.0)	2 (0.08)	10 (0.4)	10 (0.4)	35 (1.4)
G2	40 (1.6)	28 (1.1)	3 (0.12)	15 (0.6)	12 (0.5)	40 (1.6)
G3	50 (1.0)	39 (1.5)	6 (0.24)	25 (1.0)	35 (1.4)	Two 40 mm spaced 10 mm apart
S	35 (1.4)	20 (0.8)	2 (0.08)*	No results	10 (0.4)	30 (1.2)
Density in kg dm^{-3} (lb ft^{-3})						
	2.3 (145)	2.1 (130)	7.8 (485)	2.7 (170)	2.2 (140)	1.4 (85)

*Minimum practicable thickness

course other considerations than weight, aluminium might be preferred for aesthetic reasons, the GRP armour can be easily worked with standard workshop tools and can be painted and wallpapered, compressed hardwood plywood has a pleasing surface finish. GRP has the additional advantage that it is unlikely to give rise to ricochet or flying splinters.

Chapter 9

Lighting, and Power Outlets

9.1 General considerations

The principal requirement of lighting for safety and security is adequate brightness for the task in hand, whether it be work of some kind, or moving about a building, or keeping watch for intruders. By day or by night the brightness indoors is likely to be much less than the brightness of outdoor daylight. However, the eye adapts to extract the maximum visual information available in the circumstances. It rapidly contracts or relaxes the iris to adjust the size of the pupil to the level of brightness. The adjustment is slower in poor light and it adapts to night vision by activating 'rod' endings in the retina. These rods are more sensitive to light than the 'cone' endings that function in good light. The result is a marvellous range of sensitivity to light, giving some degree of vision in conditions ranging from dim starlight to intense sunlight.

For safety reasons it is important to appreciate that though the eye adapts quickly to a sudden increase in brightness, as when a light is switched on, its adaptation to a sudden decrease is much slower.

In the elderly, as described in section 4.5 the adaptation time is slower than normal. This, together with the generally poor sight of older people even in good light and the failure to interpret visual information correctly, must be provided for. Good lighting design giving absence from glare, and the modelling of shapes to produce their solid form, in addition to adequate brightness, in meeting the needs of the elderly will provide safe lighting for all users of a building.

SAFETY

9.2 Glare

Glare occurs when parts of the visual field are excessively bright in relation to the rest. Apart from the initial confusion caused by glare, the after-image can be distracting, particularly for the elderly and the visually handicapped. Glare that impairs the ability to see detail is more precisely described as *disability glare*, that which causes discomfort without directly impairing the seeing of detail is known as *discomfort glare*. Though the dangers of discomfort glare are less obvious than disability glare, like other forms of discomfort it indirectly affects safety by distraction and fatigue.

On stairways, glare can be caused by windows on intermediate landings (fig. 9.1), skylights showing the bright sky and luminaires of the artificial lighting installation directly in the user's view. Open risers allow bright lights to be seen between them, which is one of the reasons for not using open riser stairways. Highly polished wood or stone treads, also undesirable for other safety reasons, can reflect glare into the eyes of users.

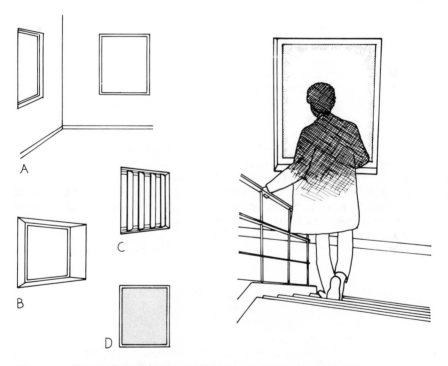

Figure 9.1 Glare from a landing window and ways of alleviating glare. (A) Additional lighting; (B) splayed reveals; (C) baffles; (D) tinted glass.

Windows at the end of corridors that are dark because of the absence of other lighting can cause glare, both directly and by reflection off highly polished floors. Where there is no alternative to positioning windows where glare is likely, the glare can be reduced by splayed reveals with light-coloured surfaces. If the amount of light admitted by a window can be reduced (perhaps its size is determined by the size of others in the elevation, rather than by the lighting requirements) then baffles can be fitted on the inside or the window can be glazed with tinted glass. Another window located on a wall at right-angles will illuminate the wall containing the troublesome window and thus reduce the contrast causing the glare. Where artificial light is used to supplement daylight, glare from windows might be reduced by increasing the interior luminance.

Security lighting makes use of disabling glare to confuse an intruder and retard his adaptation to the dark.

9.3 Stairways

The daylighting of stairways in houses is often most unsatisfactory, particularly where there is a turn in the stairs and the only source of light is a glazed front door. Of course artificial lighting is provided, but it tends not to be used. Lighting a stairway from a source above the lower landing and in front of a user as he descends will illuminate the whole of the treads and cause the user to see a highlight on the rounded nosing (fig. 9.2); this delineates the edge of the treads, assisted by the change in direction of the carpet fibres if the stairs are carpeted. If the lighting is directional relative to the edge of the treads, as from a window in a side wall or a wall luminaire, it will help in distinguishing one tread from another. Lighting that throws a high contrast shadow across stairs parallel with the treads can lead to a mis-step through a user failing to identify the edge of a tread; this is most likely at the start of a descent (fig. 9.3). Two lamps, at least, should be used to light each flight of stairways in common use, the object being to protect people from having to use an unfamilar stairway in the dark when a lamp has burnt out. The lamps should be controlled from the same switch or switches, one luminaire may contain both lamps or they may be separate.

Figure 9.2 The rounded nosings of the steps are accentuated by being lit from above the lower landing.

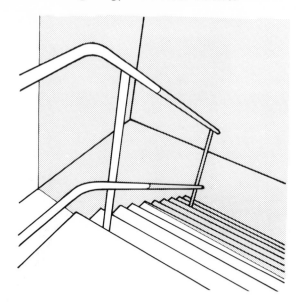

Figure 9.3 The shadow falsifies the position of the edge of the tread.

A minimum illuminance of 100 lux (9 fc) on the treads is recommended for safe movement on stairways. (Readers unfamilar with lighting design may be able to get some impression of the order of this from the knowledge that an ordinary incandescent lamp emits about 12 lumens for every watt of its rating — 1200 lm from a 100 W lamp — and 1 lumen distributed evenly over an area of 1 m² is the unit of illuminance, the lux; 1 lumen per ft² is a footcandle (fc). In houses, for the safety of elderly people in particular, the route from the bedroom to the bathroom and down the stairs must be lighted at a uniform level with dark patches eliminated.

9.4 Illuminance levels

The minimum illuminance of 100 lux (9 fc) recommended for stairways is applicable to circulating areas generally. In working areas in the home and in industry, levels of illuminance necessary for safety will generally be exceeded by the illuminance necessary to carry out tasks efficiently. The *Interior Lighting Code* of the Chartered Institute of Building Services and the Illuminating Engineering Society (U.S.A.) *Lighting Handbook* give recommendations for the lighting conditions of hundreds of different activities. Reference should be made to these codes to ensure that legal requirements for 'sufficient and suitable' lighting have been met.

9.5 Emergency lighting

Where a building will not be occupied outside the hours of daylight, and artificial emergency lighting is not installed, a minimum daylight factor of 0.1% is recommended. (Again for the benefit of readers unfamiliar with lighting design, the daylight factor is the ratio of illuminance indoors to that outdoors at the same time. In dull outdoor conditions in the U.K., outdoor illuminance is likely to be about 5000 lux. This is the value assumed in most daylight calculations. It is exceeded for about 85% of normal working time throughout the year, but may well be less when an emergency occurs.)

Artificial emergency lighting should provide an illuminance of 0.2 lux (0.02 fc) along the centre line of escape routes to meet the requirements of BS 5266, Part 1 (1975) *Code of Practice for the Emergency Lighting of Premises Other Than Cinemas and Certain Other Specified Premises Used for Entertainment.* Potential hazards such as nosings of steps, barriers and walls at right-angles to the direction of travel should be light in colour against contrasting surroundings.

Photoluminescent coatings have the property of storing light and returning it to the environment. They glow in the dark but with steadily diminishing brightness. This is compensated to some extent by the adaptation to the dark of the eye of an observer. With or without emergency lighting the coatings may be used to indicate escape routes (fig. 9.4), and mark the position of emergency equipment and controls that must be shut down during a power failure.

Figure 9.4 Use of photoluminescent coatings for emergency illuminance.

9.6   Power outlets

Safety aspects of electric and gas installations in buildings are covered in the U.K. by regulations issued by the Institution of Electrical Engineers, for wiring, and the Gas Safety Regulations made under the Gas Act for gas services. The I.E.E. *Wiring Regulations for Electrical Installations in Buildings* are accepted in England and Wales as the standard to which electrical installations should be done. In Scotland they are given legal status by the Building Standards (Scotland) Regulations. The Gas Regulations apply to England, Wales and Scotland. These and similar regulations applicable outside the U.K. must be left to be considered elsewhere, our concern here is with the number and location of outlets.

With gas, the variety of appliances that can be supplied is limited and they are usually in a fixed location, in contrast to electrical appliances. For the latter, the principal safety requirement is enough socket outlets to avoid the need for adaptor plugs. In the home it seems as though the provision will never keep up with the increasing number of appliances needing electric power. In 1974 a survey by *Handyman Which?* magazine established that subscribers would be happy with four to six outlets in the kitchen and living room, three to four in the dining room and main bedroom and one to two in the hall, landing and garage. The same people are likely to feel that more are necessary today.

A busbar track system with outlets spaced 100—200 mm (4—8 in.) apart (fig. 9.5) meets the need for multiple outlets in kitchens and living rooms. In industry it permits the safe use of plugs to continue without expensive alterations to the wiring of outlets when office layouts are changed or rooms undergo a change of use.

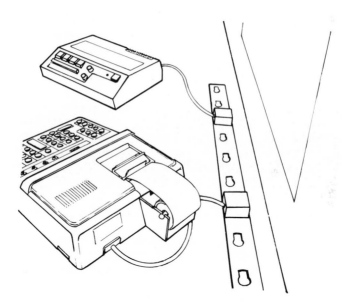

Figure 9.5 The shutter outlets of the power track system (Electrak) can be used with suitable trim and trunking on walls, skirtings, vertically on ceilings, under suspended floors and flush with floors or worktops.

In the home, socket outlets should be installed at a height of at least 300 mm (12 in.) above the floor. This will put them reasonably within stooping reach of elderly women (see table 4.2). Sockets are undoubtedly easier for everyone to reach if placed higher. Tests carried out by the Institute of Consumer Ergonomics with ninety people — fifty men and forty women of a wide range of heights, between eighteen and eighty-five years old, including seventy-four handicapped people — showed that 1 m (3 ft 3½ in.) from the floor was the one height within the comfort range for all but a few of the subjects. (This is also the most comfortable height for reaching light switches.) Unfortunately this is an example where ergonomics and aesthetics clash. Even in the interests of safety, few people wish to see socket outlets placed so much within view, and flexes hanging down the wall.

Above kitchen worktops, outlets should be at a height of 250 mm (10 in.). They must not be positioned above cookers, where the user risks an accident when reaching over the hob, and where flexes may trail across burners.

SECURITY

9.7 Security lighting

Exterior lighting removes the cover of darkness and increases the chance of a burglar being observed by natural or formal surveillance. Also, a criminal in a lighted area will have his ability to see into darker areas reduced, he will be uncertain whether he is being watched and he will be more likely to be deterred by difficulties in making an entry. The protection conferred by street lighting is shown by the way that burglaries increase in a blackout. However, because most burglaries occur in buildings that are unoccupied, there are greater benefits to be obtained from additional lighting of commercial and industrial buildings than from lighting individual residential buildings, though this is not to say that security lighting for residential buildings will not be beneficial — for something like a nurses' home it might be considered essential. Also where multi-storey blocks of flats have surroundings not covered by street lighting additional lighting is advisable: flood lighting used decoratively can enhance the appearance of the flats and the area about, give the occupants a feeling of security, and deter wrongdoing, but a 'Colditz' atmosphere must be avoided. Where school premises are lighted they benefit from the discouragement of malicious attacks during the hours of darkness.

Exterior security lighting should not leave patches of shadow, these make concealment easier for the intruder because the brighter lighted areas reduce an observer's adaptation to the dark. Preferably the illuminance should overlap so that failure of a lamp will not leave a dark area. It should not, however, cause annoyance or distraction by spreading into adjacent premises or roadways. Flat roofs should be lighted lest criminals evade detection long enough to gain access to a darkened roof where they may force an entry unobserved.

The level of illuminance required for exterior security lighting will depend on the lighting nearby. Should this be bright, main road lighting the security lighting will have to be brighter than if there were no, or little, adjacent lighting, to avoid the security lighting appearing dark by comparison. An illuminance of 1 lux (0.1 fc) may be sufficient for basic security where there is no adjacent lighting, 20—30 lux (2—3 fc) may be necessary for maximum security near main road lighting. Where the site is in a developed area, the ambient illuminance can be found with a light meter to assist the determination of a suitable design illuminance for security.

Indoors, if the building is to be patrolled by security guards it should be lighted; if guards have to use torches they will see only where they shine them — the luminance need not be bright provided it is uniform so that the guard's adaptation is not impaired. The emergency lighting system may be designed to be utilised for this purpose; additional outlets must be included to ensure that there are no dark areas in places containing items of value to an intruder.

When guards watch lighter outside areas from inside a building they need a darkened observation room with a good field of view and non-reflecting windows (fig. 9.6). Such rooms may be in a separate security building but surveillance will be less obvious if undertaken from the main building.

Figure 9.6 While the criminals are blinded by the glare, the security guard is invisible in the darkened room.

Perimeter lighting should aim at throwing glare into the eyes of an intruder while avoiding glare for the guards. If the lighting is directed against a high boundary wall, the wall should be light-coloured so that the intruder will be silhouetted against it. Lighting units should be as far inside the perimeter as practicable to give as much time as possible to detect an intruder. When a see-through perimeter fence is used, lighting directed towards the fence to cause glare will not be permissible unless the fence is in an isolated position. Where light is required to pass through chain link or similar fencing to enable an approach to the fence to be observed from inside, careful lighting design is necessary to avoid the brightness of the fence masking what lies beyond it. The colour of the mesh and of the ground surface influences the obscuring effect of the fence; a black mesh and light-coloured concrete surface provide the best visibility.

Chapter 10

Walls, Floors, Roofs and Balconies

10.1 Finishes, access and guarding

With the principal elements of a building structure, the walls, floors and roof, concern for safety lies in finishes to walls and floors, and with the guarding of roofs and other places where people might fall from a height. Concern for security lies in stopping penetration of the structure, and this includes preventing access to roofs.

SAFETY

10.2 Walls

Surfaces and arrises
Rough-textured surfaces and sharp arrises can be dangerous, particularly where children play. An example occured in a playroom intended for hyper-active children, where rough-textured bricks were used for the construction of isolated piers. Every pier presented four jagged arrises that a child could fall against or collide with. Similar examples have occurred in nursery schools.

Reflecting-glass cladding
Reflecting-glass on the exterior walls of a building may dazzle motorists, possibly to the detriment of their safe driving. There were complaints about dazzle when the building shown in F, in fig. 10.1 was erected. The elevation that is visible from a nearby roundabout is at 73° to the horizontal. Rays reflected from a surface at this angle will be horizontal when the sun is 34° above the horizon. It is then that they are likely to be most troublesome. Where reflecting-glass is used on a building with a horizontal elevation, dazzle can also occur, though over a limited distance, as shown in B. Even this limited effect should be considered in design; in winter the sun is low in the sky at a time of peak traffic in the mornings (A, fig. 10.1).

It must be noted that Local Apparent Time (L.A.T.), as shown in A, differs from the Greenwich Mean Time (G.M.T.) by which drivers

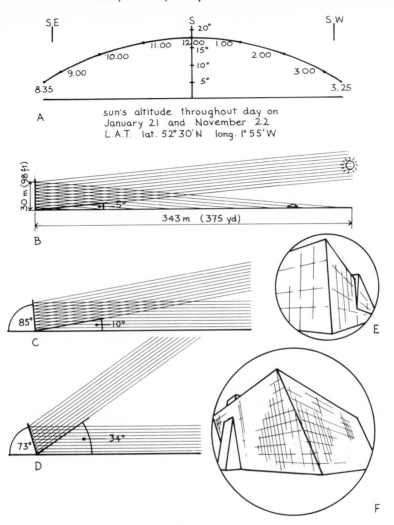

Figure 10.1 Dazzle from reflecting-glass. (A) Altitude of winter's sun at Birmingham; (B) reflection from vertical surface; (C) and (D) reflections from sloping surfaces; (E) Systime Computers building, Leeds, located where reflection poses no danger; (F) Equitable Life building, Aylesbury.

will have set their watches. For example, G.M.T. is nineteen minutes ahead of L.A.T. at Birmingham on 21 January and six minutes behind on 22 November. Thus the sun is 5° above the horizon at approximately 8.55 a.m. G.M.T. on 21 January and at 8.30 a.m. G.M.T. on 22 November. When the sun is above 5° the reflection from a vertical surface is most unlikely to be troublesome but a slight slope backwards will lead to reflection at road level when the sun is higher in the sky and will considerably increase the area affected (C, fig. 10.1).

10.3 Floors

Finishes

Accidents caused by slipping on a floor usually occur on a forward step as the rear edge of the heel meets the floor surface. At this stage of walking the other foot remains in contact with the floor, until the heel rocks forward and the leading foot is fully planted. The maximum force exerted forwards occurs shortly after the heel makes contact, it decreases as the weight of the body moves over the foot then increases to a maximum backwards as the ball of the foot starts to leave the floor. For slipping to be avoided, the friction between the sole-heel of the footwear and the floor surface must be sufficient to resist the maximum horizontal forces.

The friction between two surfaces in contact can be found by means of the classic physics teaching experiment shown in fig. 10.2.

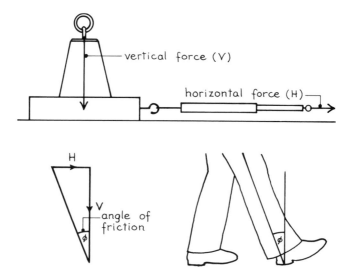

Figure 10.2 A classic friction experiment and its relationship to slipping on floor finishes.

A series of readings of the spring balance using different weights will show (within the limits of experimental error) that the ratio of (1), the horizontal force about the move the block, to (2), the vertical force pressing the two surfaces together, is constant. This constant value is the coefficient of friction (COF):

$$\text{Coefficient of friction} = \frac{\text{horizontal force } (H)}{\text{vertical force} \quad (V)}$$

This ratio is also the tangent of the angle of friction.

The heel touches the floor with the leg at an angle to the vertical as shown in fig. 10.2. It has been argued that if this angle is less than the angle of friction for particular floor finish and sole-heel materials the heel will not slip. This argument is based on the premise that only static values of friction are of importance because the heel is static in relation to the floor in walking. Another argument holds that if the dynamic COF is less than the static COF, the heel may slip on contact and not become static before a fall takes place. However, experimental work points to the conclusion that slipping is more likely to start when the foot is stationary and therefore the static COF is more relevant. (Anderson, C. and Senne, J. *eds* (1978) *Walking Surfaces: Measurement of Slip Resistance.* Philadelphia: ASTM. Perkins, P.J. 'Measurement of slip between the shoe and the ground during walking', 71—87.) In any case there is probably little difference between the static and dynamic COF when measured under realistic wear conditions. With rubber there is evidence that dynamic COF increases with velocity, thus acting against a slip with increasing efficiency.

Ratios of H/V for a limited number of subjects have been found to average 0.22 for men and 0.19 for women in normal straight walking. (Harper, F.C., Warlow, W.J. and Clarke, B.L. (1961) 'The forces applied to the floor in walking.' *National Building Studies Research Paper 32.* London: HMSO.) From these results it is suggested that a COF of 0.4 gives a high factor of safety. Nonetheless for industrial use, at least, 0.5 is considered more acceptable.

The user's contribution to the COF is largely dependent on his footwear, or lack of it. The contribution of the floor surface will be affected by wear, wet conditions and maintenance. A review by Armstrong and Lancing (Anderson, C. and Senne, J. *eds* (1978) (as above). Armstrong, P.L. and Lancing, S.G. 'Slip-resistance testing: deriving guidance from the National Electronic Surveillance System (NEISS)') of in-depth investigations of domestic accidents related to floors and flooring materials carried out by the U.S. Consumer Product Safety Commission shows that when a slipping accident occurs to someone who is merely walking about the house, usually two or more of the following factors are present:

- The floor is wet or waxed.
- The user is wearing slippery footwear such as socks, nylon stockings, leather sole shoes.
- The user is stepping from a slip-resistant covering, such as carpeting, on to waxed vinyl, or a similar surface.
- The user is making a slight change in direction.

Most of these accidents occur in the kitchen, where floors are often wet, the majority of victims are middle-aged and most are wearing shoes. Young women often slip when mopping a floor and, not surprisingly, young children slip on wet and dry surfaces when running; those who slip on dry surfaces are generally wearing socks.

Thus a floor finish for a location where wet conditions are likely must be selected for safety with that in mind. For dry conditions,

COFs obtained with a ' tester pad that simulates a rubber sole will not be relevant to circumstances where users are in stockinged feet. However, if there is to be any guidance in the selection of a safety floor finish, other than vague terms such as 'slip-proof' or 'non-slip', some sole material must be assumed. In table 10.1 the descriptions of slip resistance of floor and tread finishes are based on tests of rubber upon unpolished surfaces, and the relationship of the terms to specific COFs is indicated.

Ramps
Particular attention is required in the choice of a floor finish for a sloping surface, the alteration in the angle between the leg and the surface on which the foot is placed increases the likelihood of the heel slipping forward when the user is descending and of the toe slipping back when the user is ascending. External ramps should have a surface of roughened or grooved concrete, or brick paving with wide slip-resistant joints or another finish of at least equal traction. Note that grooves should be at an angle to the horizontal so that water does not lodge in them.

Falls on ramps often arise from the failure of users to see the edge formed by the level and sloping surfaces. The results of this failure are shown in fig. 10.3. The remedy is to make the edges as distinct as possible. Short ramps such as those often used between different levels in buildings that are altered and extended are most likely to lead to an accident. A ramp should be at least one stride

Figure 10.3 Accidents caused by failure to observe ramps. On approaching an upward slope the stride is suddenly shortened and the victim tends to fall forward. On approaching a downward slope the stride is suddenly lengthened and the victim falls backwards. Cues to the change in level are required.

Table 10.1 Slip resistance of floor and tread finishes (Reproduced from BS 5395 (1977)

Material	Slip resistance*		Remarks
	Dry and unpolished	Wet	
Clay tiles (carborundum finish)	Very good	Very good	May be suitable for external stairs
Carpet	Very good	Good	—
Clay tiles (textured)	Very good	Good	May be suitable for external stairs
Cork tiles	Very good	Good	—
PVC with non-slip granules	Very good	Good	—
PVC	Very good	Poor to fair	Slip resistance when wet may be improved if PVC is textured. Edges of sheet liable to cause tripping if not fixed firmly to base
Rubber (sheets or tiles)	Very good	Very poor	Not suitable near entrance doors
Mastic asphalt	Good	Good	—
Vinyl asbestos tiles	Good	Fair	—
Linoleum	Good	Poor to fair	Edges of sheets may cause tripping if not securely fixed to base
Concrete	Good	Poor to fair	If a textured finish or a non-slip aggregate is used, slip resistance value when wet may be increased to good
Granolithic	Good	Poor to fair	Slip resistance when wet may be improved to good by incorporating carborundum finish
Cast iron	Good	Poor to fair	Slip resistance may be acceptable when wet if open treads used
Clay tiles	Good	Poor to fair	Slip resistance when wet and polished very poor
Terrazzo	Good	Poor to fair	Non-slip nosing necessary on stairs. Slip resistance when polished or if polish is transferred by shoes from adjacent surfaces very poor

*'Very good' means surface suitable for areas where special care is required, approximates to COF > 0.75
'Good' means surface satisfactory for normal use, approximates to COF $0.4 < 0.75$
'Poor to fair' means surface below acceptable safety limits, approximates to COF 0.2 to < 0.4
'Very poor' means surface unsafe, approximates to COF < 0.2

long, say 900 mm (3 ft) to err on the right side. Handrails and balustrades are necessary to provide support and to arrest falls, as for stairways (see section 11.8).

To accommodate wheelchairs, ramps should be at least 1000 mm (3 ft 3½ in.) wide, have a maximum slope of 1 in 12 (approximately 5°) and a maximum length of 10 m (33 ft). They should be provided with a kerb to deflect wheels from the edge or from contact with a balustrade.

10.4 Roofs

Safe access

Where personnel need to have access to roofs for maintenance work on ventilation and air conditioning equipment, permanent provision for their safety should be made. If stairways cannot be provided, fixed ladders fitted with a safety cage are to be preferred. The design shown in fig. 10.4 is based on the provisions of BS 4211 (1967) *Steel Ladders for Permanent Access*, similar requirements are found in ANSI A14.3 (1982) *Safety Requirements for Fixed Ladders*.

The alternative safety provision to a cage is a rope-riding sleeve or collar to which a harness worn by the climber is attached. This should be avoided if possible. The sleeve or collar is effective in operation,

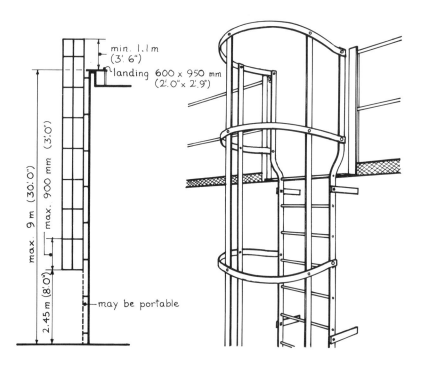

Figure 10.4 Fixed ladder with safety cage; a guard rail to the edge of an access way is also shown.

Figure 10.5 Provision for resting and tying a ladder used for occasional access to a flat roof.

it will permit free movement along the rope at normal climbing speeds, while reacting to a sudden pull caused by a fall by activating a locking or braking device which is immediately effective in stopping the fall. The drawback to its use is that a harness is seen as an encumbrance and there is always the possibility that it will not be used.

Fixed ladders are usually attached upright, but ladders at 15° to the face of the building are ergonomically better for climbing and safer in use. To avoid unauthorised use of fixed ladders the bottom section may be made detachable, alternatively the base of the ladder may be enclosed by fencing or walling and fitted with a lockable door.

Where a fixed ladder is not practicable, provision should be made for resting and tying a portable ladder (fig. 10.5). When there is no obstacle to the ladder resting directly on the building, eyebolts (see section 14.4) or other holding points should be provided.

Protection for workers
Where a roof is covered with a material insufficient to support a man's weight, the Construction Regulations 1966 require that a notice giving warning of a fragile roof covering be fixed to the building. On fragile roofs any walkway (which might be a gutter) running alongside the fragile covering must have guard rails, or alternatively there must be a protective covering on the fragile material to prevent falls through it.

If it could be construed that a walkway on a roof was used for a building operation, and cleaning a building has been held to come under that heading, then the walkway must be in accordance with the requirements of the Construction Regulations 1966. The walkway must be at least 431 mm (17 in.) wide, guard rails 0.914 to 1.143 mm (3 ft to 3 ft 9 in.) high must be provided, there must be a toe board at least 152 mm (6 in.) high and, if necessary, intermediate rails must be fitted to limit the maximum space between the toe board and the guard rail to 0.762 mm (2 ft 6 in.).

Protective barriers for building users
Balustrades and barriers at edges of roofs frequented by building users, on balconies and elsewhere that people might fall from a height should at least meet the requirements of BS 6180 (1982) *Protective Barriers in and about Buildings*, from which table 10.2 has been compiled. It is also required that designs should minimise the risk of falling, rolling or slipping through gaps in the barrier. Another requirement is that in areas frequented by children, the widest gaps in the barrier should not permit a sphere of 100 mm (4 in.) to pass through (fig. 10.6). I would, however, suggest that this requirement is always met as children get into the most unlikely places. If an exception is made for buildings where children are not permitted, the maximum gap should be one that a sphere of 300 mm (1 ft) diameter will not pass through.

SECURITY

10.5   Security aspects of walls

Surface finishes
The misuse of technological developments that has brought about the spread of graffiti on buildings is centred principally on felt-tipped pens and aerosol paint sprays, but the earlier-developed ball-point

Figure 10.6 The spacings of railings must be close enough (100 mm, or 4 in.) to remove the danger, as shown, of a small child penetrating through or passing underneath a balcony balustrade.

Table 10.2 Building user and pedestrian protection*

Barriers protecting users of:	Barriers should resist these loads separately applied			
	A horizontal UDL applied at a design level of 1100 mm	A UDL applied to the infill of:	A point load applied to any part of the infill of:	Barriers should be of a height above walk surface not less than:
	kN m^{-1} (lbf ft^{-1})	kN m^{-2} (lbf mm^{-2})	kN (lbf)	mm (in.)
Floors†, balconies, flat roofs with access, walkways and edges of sunken areas in buildings other than places of entertainment or assembly	0.74 (51)	1.0 (21)	0.50 (113)	1100 (43¼)
Floors, balconies, flat roofs with access, walkways and edges of sunken areas in places of entertainment or assembly, except that theatre balconies may be 800 mm (31½ in.) high, or less if at least 230 mm (9 in.) wide.	3.0 (206)	1.5 (31.5)	1.5 (338)	1100 (43¼)
Footways or pavements within building curtilage, adjacent to access roads, basements or sunken areas	1.0 (69)	1.0 (21)	1.0 (225)	1100 (43¼)
†Excepting floors in or serving exclusively residential buildings, where the values are:	0.36 (25)	0.5 (10.5)	0.25 (56.50)	900 (35½)

*Adapted from BS 6180

pen is also abused in this way, any surface soft enough to be indented is open to attack. Scratching with the point of a nail or penknife is another form of graffiti vandalism. Because of this and other kinds of vandalism that a soft surface is open to, only hard wall finishes should be used in a building with a potentially hostile population.

Hard surfaces are also necessary on walls likely to be attacked in public areas.

However, the felt-tipped pen and the aerosol paint spray are the bigger problem. Solutions are sought in specially formulated paints and surface treatments providing a hard, impervious coating, from which graffiti can be removed with a solvent. Multi-coloured patterned finishes are often used because they are less inviting than a light-coloured plain finish. Textured surfaces are hard to mark with a felt-tipped pen and are destructive to the tip. Unfortunately, textured surfaces are not resistant to the use of aerosols and if the surface is absorbent, like brick or concrete, cleaning is difficult and expensive, possibly requiring abrasive treatment and re-surfacing. One solution, yet to be tested by long-term use, is the application of a colourless urethane coating which seals the surface and enables it to be cleaned by commercial cleaners and solvents. As the urethane can also be used over conventional paints it offers an inexpensive means of combating graffiti in many locations. However, surface coatings on exterior masonry are generally not recommended, if one is necessary it should permit the passage of moisture and be easily removable if required. Impervious coatings lead to trouble because they do not allow the evaporation of moisture that penetrates into the wall through cracks and imperfections in the coating.

Copings

Copings that require a damp-proof course immediately under them are often easily dislodged by vandals because of the weakness of the bed joint over the damp-proof course. This weakness can be overcome by the use of clip-copings (fig. 10.7) that are grooved to fit down over the top of the wall. Like all copings, they need to be of substantial mass. Flat-bottomed copings should have sufficient mass to impose a load of at least 1.5 kN m^{-2} (32 lbf ft^{-2}), unless they are anchored down by cramps or dowels. (Concrete must have a cross sectional area of approximately 15 000 mm^2 (23 in.^{2}) to impose 1.5 kN m^{-2}.) Further resistance to dislodgement is obtained by keying units together as shown in fig. 10.7.

Wall cladding

Easily deformed, easily broken or easily removed wall cladding offers a target to the vandal: easily broken or easily removed cladding offers an access point to the burglar. The vandal can disfigure profiled

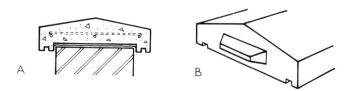

Figure 10.7 Copings resistant to dislodgement. (A) Clip coping; (B) tongued joint.

metal sheeting and siding with a slight dent and can break clay and concrete tiles used in tile hanging with a slight impact. The burglar can pull off siding or remove tile hanging to gain access to the building (fig. 10.8). He can cut out or remove the fasteners of profiled metal sheeting (fig. 10.9); it may be that the size of the sheeting is an obstacle to this sort of attack, sheets are normally available in lengths up to 10 m (33 ft), but the sort of thief who will cut a way in through an all-brick cavity wall (fig. 10.10) is likely to find 0.08 mm ($\frac{1}{32}$ in.) thick aluminium sheeting a barrier easily penetrated.

Figure 10.8 If doors and windows are well protected thieves may attempt to break in through walls by removing tile hanging or siding; plywood sheathing underneath will be an additional obstacle.

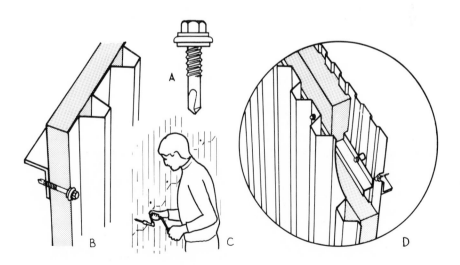

Figure 10.9 Metal cladding. (A) Self-drilling self-tapping fastener; (B) fixing of profiled sheet and insulation; (C) removing fasteners; (D) use of lining sheets with separate fixing increases security.

Figure 10.10 A bricklayer repairs a hole made by thieves to steal video recorders and television sets from a warehouse (reproduced with permission from *Express and Star*, Wolverhampton).

Not all damage to claddings is wilfully done, some will be caused by the careless or inexpert handling of ladders by window cleaners, aerial fixers and DIY decorators. To assess the resistance to damage of wall cladding, a steel ball impactor test has been devised by the Building Research Establishment (described in CP8/81 and IP 19/81). The cladding is subject to increasing amounts of impact energy until failure occurs, sufficient repeat tests being carried out to assess the percentage failure rate. The failure of brittle materials is apparent: they break. Ductile materials are dented rather than fractured, so it is necessary to define the extent of damage that is aesthetically acceptable. This will depend on position in the building, colour and texture, and probably the cost of repair. For test purposes a dent 1 mm (0.04 in.) deep as measured across a span of 50 mm (2 in.) is defined as failure. Cladding failure, as distinct from local failure, is recognised as occurring when more than 20% of the impacts cause breakage or dents as defined.

The tests show concrete tiles to have an impact resistance of 0.5 Nm (0.7 ft lbf), aluminium cladding 0.8 mm (0.03 in.) thick to have a resistance of 1Nm (1.4 ft lbf), profiled steel sheet 0.8 mm thick with plastics skin a resistance of 6 Nm (4.4 ft lbf). For comparison, window panes glazed with 4 mm thick annealed glass (various sizes and types of glazing) have a resistance of 3 Nm (2.2 ft lbf).

Based on the results of a survey of damage to buildings, the performance of claddings for various situations should meet the requirements shown in fig. 10.11. The highest requirement of 10 Nm (7.4 ft lbf) is for zones accessible to the public but not subjected to abnormally rough use. Where the risk of vandalism is high walls need to be faced with brick or material equally robust. Wall tiling and mosaic should not be used in zones accessible to the public as they showed a high incidence of damage in the survey in such situations.

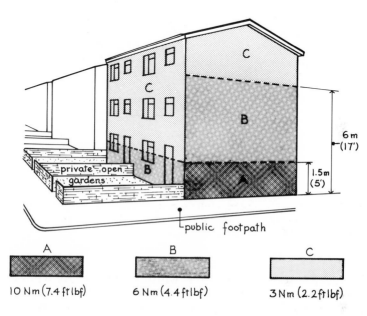

Figure 10.11 Performance levels for hard body impact resistance of wall cladding. (A) Zone readily accessible at ground level, vulnerable to hard body impact but not to abnormally rough use; (B) zone liable to impacts from thrown or kicked objects, in public locations at height or more secluded locations at low level; (C) zone not likely to be damaged by normal impacts by people or thrown objects — a lower resistance is acceptable if the cladding is easily replaceable and carefully maintained.

Party walls

Walls that are common to two adjoining buildings must form a complete vertical separation between the buildings, including the roof space, to meet fire resistance requirements. There are also sound insulation requirements. Lightweight timber framed walls with plasterboard linings are permitted in England and Wales as separating walls between houses of up to three storeys in height, provided they will give 1 hour fire resistance. In Scotland the regulations are more stringent but timber framed separating walls may be used in dwellings of not more than two storeys.

Some houses built before by-laws and regulations required adjoining houses to be completely separated have brick party walls that stop at the ceiling below the roof. These houses have been burgled by thieves breaking into one house and then making their way along the roof space to other houses, stealing what is stored in the loft and dropping down through the trap door for other plunder. The plasterboard lining to timber framed walls is not a substantial barrier to this form of housebreaking even in a double thickness. Burglars may break through in the roof space or in some situations they may just as easily break through the wall at a lower level, perhaps from an empty house that they enter in the guise of workmen or removal men. Expanded metal or reinforcement mesh securely stapled to the inside of the framing will make the wall more thief-resistant. If more than basic security is required, a masonry wall should be provided.

10.6   Security of floors

A scrutiny of the design of suspended and raised floors is necessary to see whether the floors offer the criminal the opportunity to penetrate into a building or move furtively from one part to another. Where floor panels are removable, there is a possibility that a thief may secrete items there for smuggling out later (fig. 10.12). A dedicated criminal might hide under the floor until the building is vacated at the end of the day. As the bomb under the bandstand in Regents Park that killed six bandsmen of the Royal Green Jackets in 1982 showed, terrorists can utilise a raised floor to conceal explosives. At a lower level of terrorism, arson is possible. Thus to give basic security the under-floor space of a suspended or raised floor must not be accessible to unauthorised persons. Additionally, for strong security, either the under-floor space must not be accessible from inside, or entrance and exit from the rooms where access can be obtained must be fully controlled.

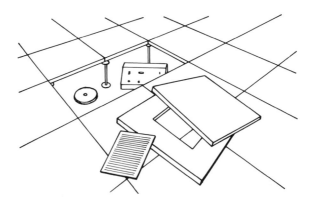

Figure 10.12 Where access can be obtained to a raised floor, stolen goods may be hidden for later removal, and there are other dangers.

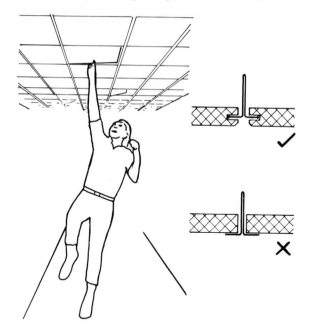

Figure 10.13 Lay-in ceiling tiles are a target for the vandal and a possible access point for the burglar.

10.7   Ceilings

Where the tiles of suspended ceilings are easily removable they provide the same opportunity for concealment of items as previously mentioned for raised floors. They may also be subject to vandalism (fig. 10.13). If the ceilings are continued over partitions, they can allow access to rooms from corridors and other rooms. Ceilings under pitched roofs can likewise become a route for the burglar.

Any soft material used for ceilings within reach of youths in schools and colleges should be avoided, once the ease with which it can be disfigured is discovered spoiling it will become a test of prowess (fig. 10.14).

10.8   Roofs and balconies

The first security requirement of roofs is the prevention of access. Climbing routes (see section 8.7) must not be created. Pipes fixed to walls externally are likely to be scaled, even if made of plastics, provided the hands can be got round the pipe to secure a grip. To prevent this, square-section pipes that fit tightly to the wall must be used (fig. 10.15). Alternatively, anti-climb guards can be fitted or the pipes can be painted with anti-climb paint (paint that remains sticky) above 2 m (6 ft 3½ in.) from the ground.

Break-ins through roofs are accomplished by removing tiles or

Figure 10.14 A vandalised, sprayed asbestos ceiling.

sheeting, or by entering through roof lights. Domelights, while resistant to breakage if made of acrylic or polycarbonate are vulnerable to entry in two ways: either the domelight can be removed from the curb, or the curb can be levered up (fig. 10.16). Clutch-head screws (see section 3.6) or other forms of security fastening will impede removal of the domelight; a secure fixing, preferably internal, will frustrate attempts to lift the curb. As a second line of defence, essential for strong security, either burglar bars or a securely held

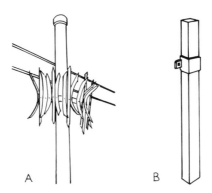

A   B

Figure 10.15 Prevention of pipe climbing. (A) Security crown (Sampson and Partners); (B) square section rainwater pipe fixed close to wall.

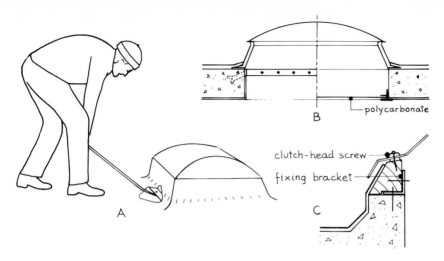

Figure 10.16 Rooflights. (A) Levering up curb; (B) burglar bars or laylight to prevent entry; (C) internal fixing for timber curb.

polycarbonate laylight can be fitted. Where these are used it is essential that the lining of the domelight and the adjacent construction be capable of providing a secure fixing.

Doors and windows accessible to burglars from balconies are more

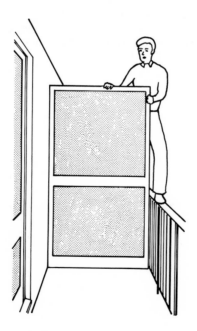

Figure 10.17 Continuous balconies are a security risk (and a safety hazard if screens are glazed with annealed glass).

likely to be left unfastened than those on a ground floor. Balconies will therefore be more secure if not continuous from one property to another. If separate balconies are not acceptable, the screen or barrier between adjoining properties must not allow passage over it or round it (fig. 10.17). For strong security, as for maximum safety, there seems no alternative but to make a balcony into a cage — or perhaps do without it.

Chapter 11

Stairways and Escalators

11.1 Stairway terms and accidents

The terms 'stair', 'stairs', 'staircase', and 'stairway' all have the same meaning nowadays. 'Stairs' is the term generally used in conversation, 'stairways' seems to be preferred for technical use. Terms used for parts of a stairway vary slightly according to the locality, those used here are illustrated in fig. 11.1.

Steps and stairs (as they are described) head the list of products most frequently involved in home accidents recorded by the Home Accident Surveillance System (HASS). In England and Wales about

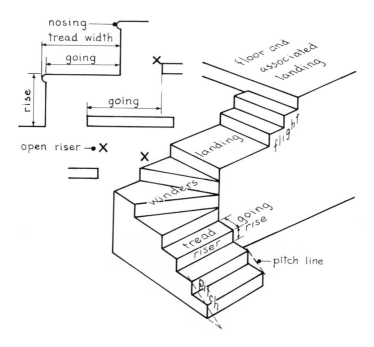

Figure 11.1 Parts of a stairway. Square-edged treads, open risers and winders are not conducive to safety.

250 000 injuries from accidents where steps or stairs have been involved are treated in hospital each year. In the U.S.A. it is estimated that between 1 800 000 and 2 660 000 disabling injuries are caused each year, and about 3 800 people die, as a result of stairway accidents.

The contribution of design to the prevention of stairway accidents lies mainly in:

- Directing attention to the presence of the stairway.
- Focusing attention on the stairs and ensuring that the steps are clearly defined.
- Providing suitably dimensioned steps.
- Providing handrails for support and assistance, and balustrades to prevent falls from the stairs.
- Providing secondary safety by eliminating items likely to exacerbate injury, and limiting distances victims can fall.
- Avoiding features likely to lead to misuse of the stairway by children.
- Avoiding increasing the hazards of decorating and maintenance above the stairs.

11.2 Directing user's attention

The need for warnings and cues to alert people so that they do not come upon stairways unexpectedly has been mentioned in section 2.4. Changes in floor surfaces, an inclination in wall decoration, an extension of the handrail (fig. 11.2), are ways in which people can be cued. Visual confusion, misleading information and distraction as might occur at the head of the stairway shown in fig. 2.4 must be

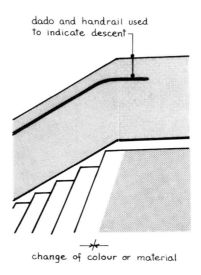

Figure 11.2 Cues to the presence of a stairway.

Figure 11.3 An indistinct stairway and a distracting scene; cues to the presence of the stairway are necessary. A ramp is the solution to the unsatisfactory step proportions.

avoided. People are particularly prone to failing to observe a stairway when the change in levels connected by the stairway is small. The dangers of one or two isolated steps are well known. A flight of more steps, outdoors, with a low pitch, and a busy street scene in front of users requires good cues to avoid accidents to people who do not see the descending stairway ahead of them (fig. 11.3). For safety a ramp is preferable in such a location.

An example of a stairway in an unexpected position behind a doorway is given in the next chapter (fig. 12.17). A landing should always be provided where a stairway runs downwards from immediately in front of a doorway. Guarding to prevent falls from the landing will be necessary, whether the stairway leads straight ahead, or to the side as shown in fig. 11.4.

11.3   Focusing attention on, and defining, steps

The user must make an appreciation of the critical parts of a stairway before using it and, once on it, concentrate on the critical parts. There must be no distraction from essential details. On an ascent the view through open risers can attract the user to look through the space between the treads instead of at the treads themselves (fig. 11.5). The eye automatically turns towards a bright light in the field of

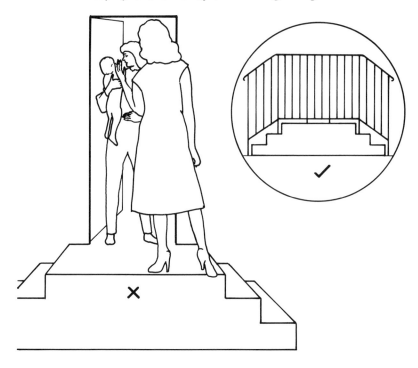

Figure 11.4 Landings need guarding even when they are only a few risers up.

view, if one is present it will be an additional stimulus to look beyond the treads. With this type of stairway it is also possible that users may suffer from vertigo induced by the view through the open risers. (The Queen's corgis had to be carried up aircraft steps with open risers until a canvas sheet was fastened under the steps to make them appear solid.) Accidents where a leg slips down between the treads appear not to happen, in spite of the seemingly obvious danger.

The brightness and flickering of a television set visible from a stairway is a strong attraction (fig. 11.6), taking the user's attention away from the stairway. Similar distractions are a great number of changes in the user's visual surroundings, and rich scenes suddenly presented as the user's view passes below the horizontal edge of the ceiling, as shown in fig. 11.7, or passes by a vertical edge at the side of a flight. These edges are known as orientation edges because they cause the user to attempt to orient towards the scene and away from the stairway.

Colour and lighting (see sections 9.2 and 9.3) should be used to define steps and handrails clearly, so that they are the most prominent objects in the user's visual field. (Stairways built entirely of transparent plastics are not a good idea.) The saw-tooth shape of the steps should stand out clearly. Lines parallel with the edges of treads, camouflaging and busy patterns must not be permitted (figs 11.8 — 11.10). Shadows must be avoided, particularly if they are parallel with the treads (see fig. 9.3). Making the steps the outstanding feature is most important when a rich view from the stairway is unavoidable.

11.4 Proportions of steps

In about 1672 François Blondel, director of the Royal Academy of Architecture in Paris, concluded that the rise and going of a step should be related to the average human pace. He took this as twenty-four inches; he also recognised that pace shortened on departure from level walking. People were less tall then and the inch was more than 25.4 mm but the formula Blondel devised — twice the rise plus the going equals twenty-four inches ($2r + g = 24$) is still the standard method of determining the proportions of steps today, though the

Figure 11.5 Open risers allow distracting views on ascent, users may suffer from vertigo, and the treads are difficult to carpet. Other safety hazards of this professionally designed stairway, as well as the alarming absence of guarding, are: handrail not returned to wall; square-edged treads, probably slippery; wired glass screen close to bottom (reproduced with permission from *The Architects' Journal*).

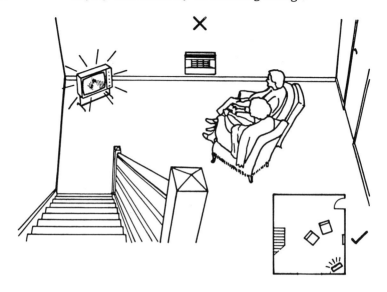

Figure 11.6 If a stairway must be in a habitable room, planning should encourage occupants to place the television set out of sight of people descending.

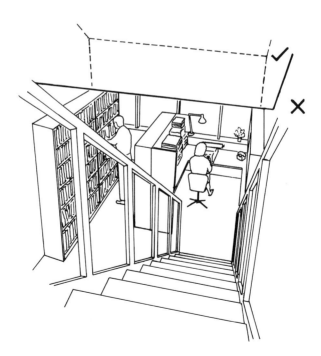

Figure 11.7 When a stairway user passes under this orientation edge the rich view suddenly attracts his attention.

Figure 11.8 The exposed aggregate finish makes the edge of the treads difficult to distinguish especially by an elderly person seeing them slightly out of focus, as at the top of this picture.

twenty-four inches must not be assumed to be precise. The requirements of the building regulations of England and Wales are met if $2r + g$ lies between 550 and 700 mm ($21\frac{5}{8}$ and $27\frac{1}{2}$ in.). BS 5395 (1977) *Stairs, Ladders and Walkways* and other building codes have similar requirements. However, Blondel's formula applies only over a limited range of dimensions. Steep flights can have a going that requires the person descending to twist sideways to get a ball of the foot on the tread, upsetting balance and increasing the likelihood of an accident, and yet conform to the formula. For this reason a minimum going or maximum pitch (or both) is imposed. Unsuitable flat-pitched flights that impose an awkward gait on the user (fig. 11.11) may be prevented by the requirement of a minimum rise or minimum pitch.

Figure 11.9 These treads would be more distinct if they had rounded nosings, they would also be less likely to exacerbate injury when a fall occurs.

In Britain, for residential buildings, the maximum pitch considered acceptable for stairways serving only one dwelling is 42°; for stairways in common use in connection with two or more dwellings the maximum pitch is 38°. There are different ideas about the pitch that is ergonomically preferable for stairways generally, but the consensus of opinion favours 30°. Minimum pitches are seldom specified.

In the U.S.A., after experiments with an adjustable mechanical stairway, Fitch, Templer and Corcoran (Fitch, J.M., Templer, J. and Corcoran, P. (1974) 'The dimensions of stairs'. *Scientific American*, 231, 82—90) concluded that a linear equation such as Blondel's was unresponsive to human adaptability and that safety design in terms of a low rate of mis-steps was obtainable from a range of combinations using a rise of 4—7 in. coupled with a going of 11—14 in.

Figure 11.10 Crazy paving, the ultimate in camouflaging patterns for stairways; carpeting with a 'busy' pattern can produce a similar effect.

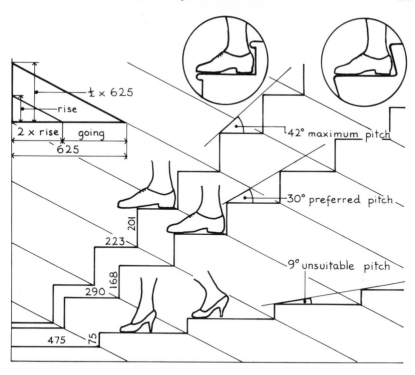

Figure 11.11 Stairway pitches when twice the rise plus the going equals 625 mm (24½ in.): the mid-point of the range 550 — 700 mm (21⅝ — 27½ in.). The man's foot shown is of the size of the ninety-fifth percentile with normal footwear. The woman's feet show that she does not move forward a full going with each stride on the shallow pitch. The insert shows how a nosing or angled riser provides extra width for the foot on descent.

(in S.I. units these dimensions can be taken as 100—180 mm and 280—355 mm). Fitch and his colleagues were attracted to the subject of stairway safety by reports of a high accident rate on a monumental stairway of four broad steps leading to a walkway in front of the Metropolitan Opera House in New York. It appeared that people stumbled on the stairway because of the awkward gait imposed on them by a rise of 86 mm (3⅛ in.) and a going of 635 mm (25 in.).

Using the smallest rise and the largest going of Fitch's recommendations to obtain the flattest pitch gives 16°. The steepest pitch obtained with the largest rise and smallest going is 32°. This is about the steepest advisable for places of public assembly, hospitals, old people's homes and similar buildings (the Building Standards (Scotland) Regulations 1981 state a maximum of 33°).

Fitch's range of dimensions for rise and going are recommended in 'Guidelines for stair safety' (*N.B.S. Building Science Series 120*) which was prepared for the U.S. Consumer Product Safety Commission. Unfortunately, for domestic stairways (in Britain at least) the constraints of space are likely to demand pitches steeper than those within this range. In the absence of further advice, recourse must be had to the $2r + g$ type of formula, with a minimum going of 240 mm (9½ in.) for these stairways. Mostly an easy going is preferable to an easy rise for the benefit of the elderly to rest without fear of overbalancing. For children, an easy rise is more suitable; but we should give preference to old people on account of their greater vulnerability to injury and reduced powers of recovery.

Where a step has an overhanging nosing, the width of tread will be greater than the going, as shown in fig. 11.11. Some authorities specify a minimum width of tread as distinct from, and sometimes as well as, a minimum going. But the extra tread width is only advantageous if the bulge of the heel utilises the space under the nosing on descent. Perhaps up to 15 mm (⅝ in.) of the overhang of the nosing (or of one tread over another on open-riser steps) can be taken up by the heel but any greater overhang should be disregarded. The recommended minimum tread width based on the length of the ninety-fifth percentile male foot (see table 4.2) is 298 mm (11¾ in.), this is consistent with the smallest going of 280 mm (11 in.) of Fitch's recommendations.

Winders
Uniformity in the rise and going of steps is necessary because the stairway user quickly gets into a rhythm of foot movements. On ascent the foot is lifted only a few millimetres above the treads, a variation of 5 or 6 mm in the rise of steps can cause a user to stumble. Similar small irregularities in the going can also cause the foot to be misplaced. It follows that winders in a stairway are undesirable because their going is different to that of ordinary treads (flyers). Additionally they are not of adequate width for the proper placement of the foot at the newel end.

As winders taper towards the newel the pitch becomes steeper until, at the newel, it is practically vertical. In falling a victim can slide round the corner and be subjected to a considerable drop. The outer handrail necessarily stops at the newel so the user has three winders to negotiate without a handrail for support; it will be six winders if a half space of winders is used (fig. 11.12). A handrail must be fixed to the wall at the other side of the stairway, but where the handrail runs into the corner it will be a long reach from the walking line. The lofty ceiling over the winders coupled with the difficulty of working over them can make re-decoration hazardous for the D.I.Y. decorator, as shown in fig. 11.12.

11.5   Spiral and helical stairways

It is said that spiral stairways (fig. 11.13) are safe because everyone knows that they are unsafe — which should be enough to condemn them, though nothing much is known about accidents on this type of stairway. The treads are at least uniform, though of course too narrow at the centre of the spiral. Here the pitch approaches the

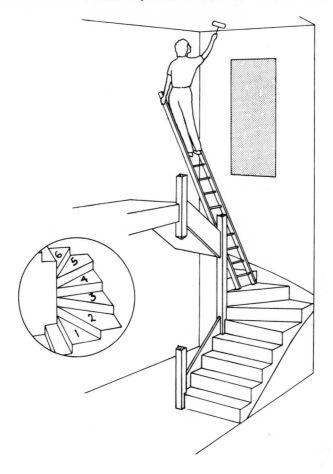

Figure 11.12 Some hazards of winders: their steepness (see insert); the difficulty of decorating; the difficulty of reaching the window.

vertical and a user who slips may fall heavily. There are difficulties in obtaining sufficient headroom between the steps and landings, and there is much danger of people colliding with the steps at floor level (fig. 11.14).

Helical stairways, where the tapered steps curve round a central void or curved wall instead of the column of spiral stairways, are subject to the same objections. The steps do not taper as narrowly, but as these stairways are usually made wide enough for people on them to pass each other, users are likely to be forced to alter their gait as they move across the steps in avoiding each other. This puts them at risk of a fall.

Stairways that curve round a central wall may also be a security risk. Muggers can choose a position near the top or bottom where they can lurk out of sight of prospective victims and are able to make

Figure 11.13 The spiral stairway has two handrails but as it is in a library most users will have one hand occupied and therefore be unable to take advantage of this safety feature. Also the surroundings distract the user's attention from the stairway. The climbable balustrade of the balcony would be unsuitable in a library where children accompany their parents (reproduced with permission from *The Architects' Journal*).

Figure 11.14 Low headroom — one of the dangers of a spiral stairway.

a quick and hidden getaway in a manner that is not possible with straight flights.

Recommendations for the design of helical and spiral stairways which will make them as safe as possible are given in BS 5395, Part 2 (1984) *Stairs, Ladders and Walkways.*

11.6 Headroom

The headroom requirement of 2000 mm (6 ft 7 in.) based on the clearance needed for men age 25—34 wearing shoes with 25 mm (1 in.) high heels, given in table 4.2, is generally acceptable in Britain. The preferred clearance in the U.S.A. is 2100 mm (6 ft 10½ in.) with a minimum of 2030 mm (6 ft 8 in.). Headroom above a flight is measured vertically from the pitch line (fig. 11.1). Above a floor, headroom should not be less than that to be maintained above flights, otherwise collision with the underside of the steps is likely to occur through children running about, as shown in fig. 11.14, and through people walking into the soffit when not watching where they are going, as shown in fig. 11.15.

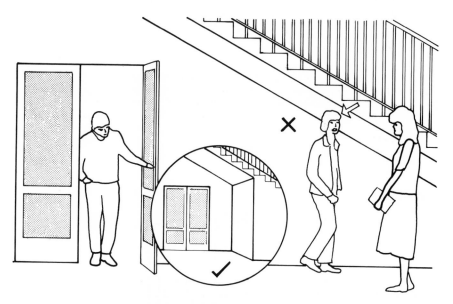

Figure 11.15 Another low headroom hazard.

11.7 Slip resistance

Slipping on a stairway is almost impossible if treads are dry and clean and the foot is placed correctly: the horizontal component of the load transmitted through the foot is much less than with floors (section 10.3). However treads get wet, directly if they are exposed to the weather and indirectly through users bringing water indoors on their feet. Friction is reduced and if the steps are very wet hydroplaning can occur. People wear shoes with slippery soles or go about in their stockinged feet. When hurrying down stairs they do not place their feet carefully (fig. 11.16) with the result that the surface in contact with the tread is considerably reduced.

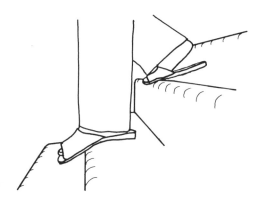

Figure 11.16 Reduced contact between tread and foot is the usual cause of slipping.

To guard against slipping, externally, treads should have a perforated or well-drained surface, internally, polished wood, polished stone and smooth ceramic tiles should not be used; mat wells should be provided so that mats can be used to remove moisture from the feet of people entering the building. Non-slip strips parallel with the edge of the tread should not be used because they are likely to confuse a user with bad eyesight as to which of the lines he sees is the edge of the tread. Friction must not be too great or the foot may lock on the tread surfaces. A user needs to be able to pivot and slide his foot slightly, in turning to avoid someone else in the stairway, for example. For this reason rubber matting is not a suitable covering for treads. Research findings point to a static coefficient of friction of 0.6 as preferable. Any applied items such as adhesive abrasive strips and applied nosing pieces should be avoided, because of the possibility of separation after prolonged use. Further guidance on the suitability of tread finishes is given in table 10.1 in the previous chapter.

11.8   Handrails and balustrades

When a loss of balance occurs on a stairway the reflex grasping action of the hand needs to be accommodated by a conveniently placed rail that can be gripped by the hand with the thumb and index finger forming the shape of the letter C. Tubular rails of steel and aluminium are suitable, but wood handrails often fail to meet the shape requirement. The design of wood handrails has declined over the years, they used to be well-rounded and of a size that could be grasped comfortably, now they come in the form of cappings to balustrades, too wide to afford a grip, or narrow boards on edge that have to be held by the hand rather than in the hand (fig. 11.17). A smooth rounded surface to a handrail enables the hand to be slid along to monitor progress and maintain stability. The elderly require a handrail they can grasp comfortably for support and assistance. A diameter of 45—50 mm (1¾—2 in.) is recommended, with a maximum width for a shaped rail of 65 mm (2⅝ in.).

The handrail should be uninterrupted for the length of the stairway and continue horizontally for about 300 mm (1 ft) at the head of the stairway and about 300 mm plus a tread width at the foot of the stairway, as shown in fig. 11.17, to lead the user into and out of the stairway. Falls from the bottom step often occur when a user mistakenly thinks the floor level has been reached (fig. 11.18), the break to the horizontal on a handrail can be the cue the habitual user of the stairway subconsciously waits for on descent.

For gripping in an emergency the handrail needs to be at least 65 mm (2⅝ in.) clear of the wall. The ends of the handrail should be returned to the wall so that sleeves, handbags, etc., are not caught and the stairway user thrown off balance.

Within limits, users can choose their own height of the handrail to use by adjusting the distance in front of their body that they grasp the handrail. The generally accepted range of heights above the pitch line of 840—1000 mm (2 ft 9 in.—3 ft 3½ in.) is satisfactory

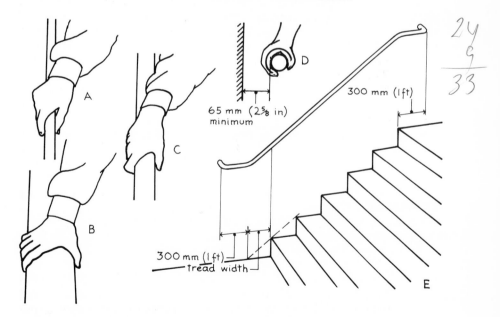

Figure 11.17 Handrails. (A) Too narrow; (B) too wide; (C) just right; (D) providing C-shaped grip well clear of wall surface; (E) finish at head and foot of stairway.

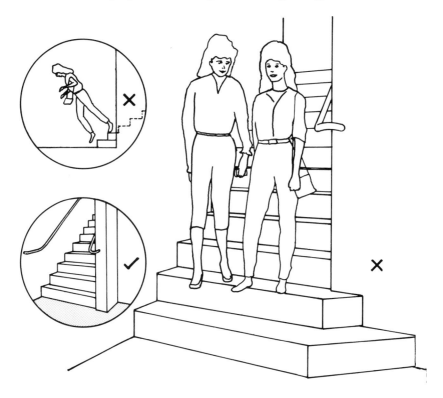

Figure 11.18 On arriving at the vertical wall edge, the user of the stairway
who is not paying full attention thinks the end of the flight has been reached:
the end of the flight should be unambiguously indicated.

for adults. To cater for children, a rail about 610 mm (2 ft) above
the pitch line should be provided, this should not be more than
38 mm (1½ in.) wide. For the elderly, and for safety generally,
handrails each side of a user are necessary (fig. 11.19). These may
be fixed each side of the flight if the width permits only one person
on the flight at a time. If the flight is wider, intermediate handrails
should be provided so that there is one within arm's length of every
user. However, there should be at least 1000 mm (3 ft 3 in.) clear
between the handrails.

Balustrades guarding stairways are commonly topped by a handrail,
thus handrail height becomes balustrade height, subject to the
requirements described below. This is generally satisfactory for
safety provided the balustrade cannot be climbed by children
(fig. 11.20) and provided the handrail does not have to be of excessive
width. Traditionally, balustrades on landings are made half a riser
higher than the height, measured vertically, of balustrades on flights.
BS 6180 (1982) *Protective Barriers in and about Buildings* specifies
balustrade heights of 850 mm (2 ft 9½ in.) for flights and 900 mm
(2 ft 11½ in.) for landings in residential buildings. In other buildings,

900 mm above the flights and 1100 mm (3 ft 7¼ in.) above landings are specified. Based on the ninety-ninth percentile of the centre of gravity of men, age 25—34 (see table 4.2), a height of 1150 mm (3 ft 9¼ in.) is advisable.

In buildings where children under five live, there should be no openings in a balustrade through which a sphere of 100 mm (4 in.) can pass. However, children enter buildings of all types and like to explore whenever they get the opportunity, therefore this safety precaution should be extended to other buildings. For industrial buildings and elsewhere, where children are not allowed, a maximum spacing of balustrade members of 500 mm (19½ in.) is sometimes recommended, but an adult could slip through rails at this spacing on a flight or landing (fig. 11.21). Another example of unsatisfactory spacing of balustrade members and another danger brought about by inadequate guarding is shown in fig. 11.22.

Loads that balustrades for ramps, landings and floors should be designed to resist, separately applied, in (a) residential buildings and (b) other buildings, except places of entertainment and assembly, are given in BS 6180 as follows:

- A horizontal UDL applied at the design level of 1100 mm (3 ft 7½ in.) of
  (a) 0.36 kN m^{-1} (25 lbf ft^{-1})
  (b) 0.74 kN m^{-1} (51 lbf ft^{-1})

Figure 11.19 Safe stairways need double handrails for the elderly, low level handrails for children.

- A UDL applied to the infill of
  (a) 0.5 kN m^{-2} (10.5 lbf ft^{-2})
  (b) 1.0 kN m^{-2} (21 lbf ft^{-2})
- A point load applied to any part of the infill of
  (a) 0.25 kN (56.5 lbf)
  (b) 0.50 kN (113 lbf)

In places of entertainment and assembly, such as theatres, cinemas, concert halls, assembly halls and stadia, to meet the requirements of BS 6180, stairway balustrades should be at least 900 mm (3 ft 11½ in.) high and resist a horizontal load of 3.0 kN m^{-1} (206 lbf ft^{-1}) at the design level of 1100 mm (3 ft 7¼ in.). The infill should resist a UDL of 1.5 kN m^{-2} (31.5 lbf ft^{-2}) and a point load of 1.5 kN (337.5 lbf).

Figure 11.20 A type of balustrade that children can climb, or get through when the spaces between the members are larger. The narrow sharp boards are unsatisfactory as handrails, however, the lower rails are better than nothing for children. Note that there will be difficulty in cleaning the window in safety.

Figure 11.21 A balustrade through which an adult could fall. A maximum spacing of 300 mm (1 ft) is recommended. Note the gap by the wall and see insert below.

11.9 Doorways and windows

In systematic safety design, the siting of doorways and windows in relation to stairways must be considered along with the design of the stairway. The reader is therefore referred to section 12.4 for dangers arising from the proximity of doorways to stairways and to sections 14.1 and 14.2 for some dangers arising from windows on stairways. The hazards associated with reaching windows over winders are illustrated in fig. 11.12. Windows over other parts of

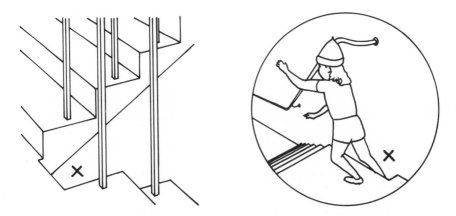

Figure 11.22 These dangers are increased when the stairways are crowded.

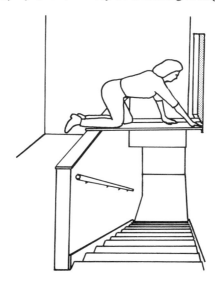

Figure 11.23 How one housewife solved the problem of reaching a window over a stairway; the boards were kept behind a wardrobe when not in use.

stairways can also cause householders to put themselves in jeopardy when they fit or change curtains, clean the glass or decorate the window. The danger one householder was put in is shown in fig. 11.23. Windows placed as shown in fig. 11.20 also tempt people into taking risks.

Lighting stairways from windows and luminaires is covered in sections 9.2 and 9.3.

11.10 Secondary safety

To minimise injury when a fall occurs on a stairway, every effort should be made to eliminate sharp edges from the design. This is another reason for using rounded nosings; brick steps with sharp, jagged edges will cause painful cuts and lacerations, the same bricks with rounded edges would probably confine the injury to bruising, if that. When a person falls on ascending a stairway, the reflex response is to put out the hands in an attempt to break the fall. If the hands contact a sharp edge, cuts and lacerations are likely, whereas with a rounded nosing no harm might be done. With open risers if the hands miss a tread the victim might fall forward and his face impact with the edge of a tread.

Restricting the number of risers in a flight will limit the distance a victim falls, a maximum of sixteen is a common recommendation, but fewer are preferable.

11.11 Escalators

Falls on boarding or leaving are the most common accidents on escalators, but accidents involving the entrapment of footwear cause the most public concern. Provisions aimed at reducing both these types of accident are among the requirements of BS 5656 (1983) *Safety Rules for the Construction and Installation of Escalators and Passenger Conveyors*, which is identical with the European Standard EN 115.

After a study of falls on escalators in the U.S.A. by W.F. Zimmerman and G.K. Deshpande ((1978) 'Escalator design features evaluation', *J.P.L. Publication 81*, California Institute of Technology) and consideration of the time taken by elderly people to react, the researchers concluded that a minimum of 1½ or 2 horizontal running steps, according to the speed that the steps move, with a preference for 3 or 3½, should be provided at top and bottom landings (fig. 11.24). BS 5656 puts the requirement at a minimum of 0.80 m (2 ft 8 in.) horizontal travel for rises of escalators not exceeding 6 m (19 ft 9 in.) and speeds not exceeding 0.50 m s⁻¹ (100 ft min⁻¹), otherwise the minimum horizontal travel should be at least 1.2 m (4 ft).

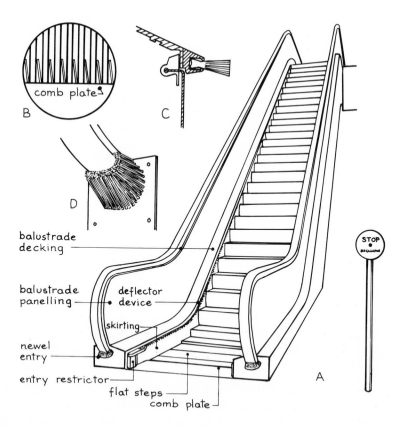

Figure 11.24 (A) Parts of an escalator; (B) trip switches actuated by foreign bodies trapped between grooves of step at run-in to comb plate; (C) brush bristle deflector device (O and K Escalators); (D) entrapment protection at newel entry.

The provision of more horizontally running steps than required under the previous British Standard gives passengers time to adjust their balance, on boarding, before the steps separate and to gain their balance, before leaving.

To avoid congestion at entrance and exit and to give passengers time to prepare for boarding and recover from leaving, the depth of the landing should be at least 2 m (6 ft 7 in.) from the end of the handrail. Should the width of the landing be less than twice the width of the escalator, a depth of at least 2.5 m (8 ft 3 in.) is necessary. Consideration will need to be given to increased depth where the landing is part of a general traffic route.

Entrapment of footwear involves young children who are wearing plastic wellington boots or trainer type shoes; any soft footwear is potentially vulnerable unless precautions are taken, and Zimmerman and Deshpande warn against wet tennis shoes in particular. Serious injury has occurred when the leg portion of a child's wellington boot has been drawn into the gap between the edge of a step riser and the skirting at the start of downward travel. The accident begins with some drag on the boot when it is in contact with the skirting. Then, as the steps separate, the relative upward motion of the riser of the step behind the step upon which the foot is placed, and the relative backward motion of the skirting, draw the leg of the boot into the gap. D.H. Weston, in his investigations into this type of accident (technical information supplied by O and K Escalators), has shown that the time taken for the boot to start exerting a tourniquet effect on the child's leg is probably less than 1½ seconds from when the steps start to separate.

The function of the entry restrictor and deflector device shown in fig. 11.24 is to keep feet from contact with the skirting and away from the gap at the end of the steps. Such accident prevention items may be fitted to existing escalators. It is important that their design does not create the further danger of feet being trapped under the restrictor or deflector itself.

In the U.S.A., reliance is put on stopping the escalator when entrapment occurs. The ANSI standard A17.1 (1982) requires the skirting to deflect and trip an emergency switch, but Weston has pointed out that having a skirting sensitive enough to react to the sort of entrapment described above may result in stoppages by a normal user inadvertently pressing a foot against the skirting. If entrapment has occurred the flexible skirting will deflect just at the time the clearance should be kept to a minimum. In any case, an escalator cannot be allowed to make an instantaneous stop — this would throw off other passengers — thus in the one second approximately in which it can stop with reasonable safety to others, the victim could already have sustained injury by further drawing in of the footwear.

To minimise the effect of trapping of objects or clothing at the combplates, trip switches are fitted as shown in B, in fig. 11.24. A trip switch is also fitted at the entry point of the handrail to the newel, in case some foreign body gets past the brush guard (D).

Trapping under the handrail should not occur, provided regular maintenance is carried out to ensure that the handrail does not become slack by stretching.

Emergency stopping switches for use by the public when necessary must be provided on landings. The requirement of the American standard A17.1 (1982) for emergency stop buttons to be located in the right hand newel base facing the escalators on both landings is in sharp contrast to what the Health and Safety Executive sees as safe practice in the U.K. In Guidance Note PM 34, *Safety in the Use of Escalators*, the H.S.E. says that emergency stop devices should not be placed where they are difficult to find 'e.g. mounted on skirtings and newels'. They wish to see emergency stop devices in prominent positions, preferably out of the reach of children. The American standard permits emergency stop buttons to have unlocked covers over them; BS 5656 requires emergency stop devices to be conspicuous, easily accessible, coloured red, and marked 'STOP'.

Notices advising passengers on the use of escalators are required by BS 5656, by the Building Standards (Scotland) Regulations 1981 and by A17.1, and these are to be placed near the entrance. It can only be hoped that these are read and understood by the public before they use the escalator, as at the time of approaching the entrance passengers need to focus full attention on to the act of boarding.

Chapter 12

Doors

12.1 Door accidents

Doors are one of the few moving parts of building construction, so it is not surprising that they figure prominently in accidents in buildings. The Home Accident Surveillance System (H.A.S.S.) shows that doors are the third most frequently involved product in all home accidents. Children are particularly susceptible to door accidents, the up to 15 age group (0—14) comprises only 22% of the population of England and Wales, yet it has 56% of door accidents in domestic buildings. Regrettably, there are no statistics available about door accidents in other types of buildings.

Data obtained by the H.A.S.S. are used for the analysis given in table 12.1, this shows that there is no significant difference in the type of door involved with either children or adults. As far as can be deduced from the data, children simply suffer more of the same. Glazed storm doors constitute an additional hazard in the U.S.A. and in other countries where their use is common.

Because a door moves, accidents involving doors fall into the *struck-by* category as well as the more common *struck-against* category of construction items. *Caught-in* and *caught-by* accidents constitute the third and fourth categories of door accidents. Caught-in accidents involve the door and the frame, as when fingers are caught in between the pivoted side of a swing door and the adjacent frame. Caught-by accidents involve door furniture. An appraisal of the safety of a doorway (using the checklist in fig. 2.5) should anticipate each of these kinds of accident. The appraisal must also include the possibility of an accident where the door itself is not involved: the kind of accident that results from the siting of the doorway.

12.2 Glazed doors

Struck-by and struck-against accidents involving a glazed door, which would scarcely cause comment if a safe material had been used, can

Table 12.1 Door accidents in the home

Category	All ages		0—14 age group		
	No.	% of category	No.	% of category	% of 0—14 total
Door involving glass	857	26	456	53	25
Door, all other	2327	71	1347	58	72
Door furniture	113	3	56	50	3
	3297		1859		

$$\frac{0\text{—}14}{\text{All ages total}} = \frac{1859}{3297} = 56\%$$

be fatal if the door contains annealed glass. Probably nearly 20% of accidents involving glass in doors occur through pushing or striking to open or close the door; Webber and Clark found 18.6% of 365 accidents with glazed doors occurred in this way. (Webber, G.M.B. and Clark, A.J. (1981) 'Accidents involving glass in domestic doors and window: some implications for design'. *Building Research Establishment Information Paper IP 18/81*.) Victims may push on the glass deliberately, or reach for the handle or push-plate as they approach, and strike the glass accidentally (fig. 12.1).

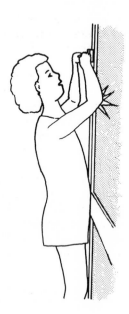

Figure 12.1 The striking or pushing of the glass in a door that leads to breakage may be indirectly applied, as in the case of a child attempting to unlock a night latch almost beyond her reach.

Figure 12.2 Disputed access is a common cause of door accidents to children.

People are cut by glass as doors are closed in front of them (perhaps deliberately), either when they are following closely behind someone or when they are approaching from the other direction (fig. 12.2).

Loss of balance is another common cause of accident with glazed doors. Injuries are usually more serious than those inflicted when opening or closing doors because a larger part of the body is involved (fig. 12.3). Most loss-of-balance accidents happen when people slip

Figure 12.3 A fatal fall occurred when a child lost balance while doing a hand-stand.

Figure 12.4 Children chase each other in play particularly along passageways. When a child falls, or in haste reaches for the door handle and misses, the glass can be struck with a great deal of force and serious injury be caused.

or stumble, sometimes people are pushed. Often the victim falls on or from steps or stairs. Children fall while at play (fig. 12.4).

Doors may also be slammed dangerously by the wind. In ordinary houses housewives have been injured as their glazed front doors slammed on them as they collected the milk from the doorstep. In tall blocks of flats, where the wind has to speed up as it gets round the obstruction presented to its progress and attempts, as it were, to catch up with the rest of the wind, it will readily take a path through the building if one is available to it. A clear path may require swing doors to be opened, and this the wind will do. The doors get blown about and strike people entering or leaving. Ground floor corridors with glazed swing doors at each end can be particularly troublesome. The example shown in fig. 12.5 is an old people's home, not a high rise building, but one with a vulnerable population.

Forgetfulness, lack of perception and distraction lead to accidents where a person walks into a glazed door, not realising it is there or that the door is closed (fig. 12.6). Sliding doors with a large area of glazing and narrow framing, such as 'patio doors' are conducive to this sort of accident. In one accident with a patio door a householder came in from the garden as the evening drew in, washed his hands, then noticed that he had left the lawn sprinkler on. He hurried across the darkening room to turn off the sprinkler but forgot he had closed the door on the way in. The severe cuts and lacerations he suffered kept him off work for several weeks — he might have lost his life.

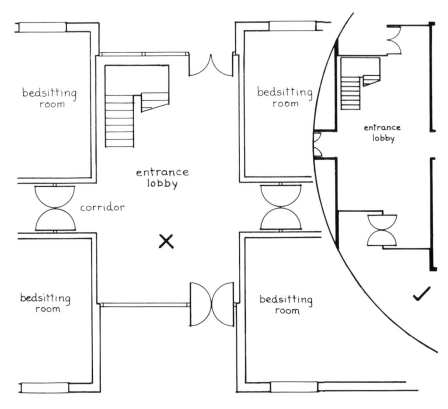

Figure 12.5 An entrance lobby that becomes a wind tunnel, creating dangers with the doors. The insert shows how the problem may be alleviated.

The simple solution for doors that must contain translucent material is the use of safety glazing material. Polycarbonate or other plastics sheet can be used only in certain situations; laminated glass is as fragile as annealed glass, though safer, and toughened glass is still breakable and can cause lacerations (see chapter 6), so the siting of the doorway and the location of glazing in the door require careful consideration. Perhaps the door can be re-located, and perhaps a translucent material is not necessary after all.

The British Standard *Code of Practice for Glazing for Buildings*, BS 6262 (1982) specifies safety glazing material for fully glazed doors. However the definition of a fully glazed door in the code is one where the glazing comes within 150 mm (6 in.) of either edge and the top of the door *and* within 300 mm (1 ft) of the bottom, hence if the framing exceeds 150 mm width at the top or 300 mm at the bottom the door is not considered to be *fully* glazed and annealed glass may be used. Annealed glass may also be used in doors with mid-rails and in fully glazed doors fitted with a protective rail or rails independent of the glazing. The rails have to be at a height of 800 mm to 1100 mm (31½ to 43¼ in.) above the floor, and have a combined maximum width of 75 mm (3 in.) on each side of the door.

The Child Accident Prevention Trust (C.A.P.T.) does not accept that BS 6262 provides adequately for the safety of children. It believes that annealed glass can only be used with reasonable safety in doors when it does not extend below 1500 mm from the floor level; the only exception being a small area (0.05 m², or 77 in.²) of a vision panel. This is allowed to reach below 1500 mm from the floor, providing it is not within 200 mm of the swinging edge of the door.

As children are likely to be found at some time in any building the C.A.P.T. recommendations should always be followed — they also, of course, provide for the greater safety of adults.

Side lights and other glazing that might be mistaken for a door must be subject to the same safety considerations as doors.

12.3 Swing doors

The first safety requirement of a swing door is that it should allow good forward vision of what lies on the other side to anyone approaching the door (fig. 12.7). The shortest user should be able to see through and be seen through the door; a transparent panel is therefore essential.

An eighteen-month-old child is the shortest person who might be walking towards, or standing by, a swing door. A child of this age could wander away from an inattentive adult to explore the

Figure 12.6 A boy was killed when he ran to tell his brother to come and see something on television.

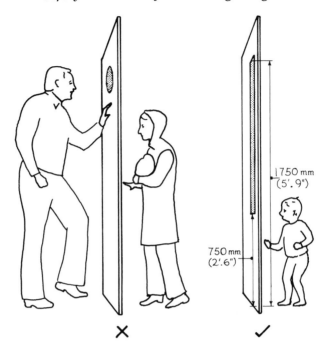

Figure 12.7 Users must have a clear view ahead of swing doors.

surroundings. To cater for this, a vision panel should reach to within 750 mm (2ft 6 in.) from the floor.

A person approaching the door will be looking downwards and forward, so the top of the vision panel may, with safety, be below his eye level. However he will not feel comfortable unless he can see more-or-less straight ahead. A minimum height of 1750 mm (5 ft 9 in.) is suggested. If wired glass must be used for fire safety then the wired glass should be one component of a laminated glass panel, alternatively the wired glass should be made shatter-resistant by bonding on polyester film.

With frameless glass doors and aluminium doors, fingers can be trapped between the edge of the door and the jamb or mullion of the frame or shopfront (fig. 12.8). As the door swings, a gap, perhaps 50 to 75 mm (2 to 3 in.) wide opens up. The thinness and weight of glass doors makes them the most dangerous, but the leverage exerted by any door as it closes brings considerable force to bear on fingers inadvertently placed in the gap. The pivots of wood doors can be positioned close to the heel, so that the gap that opens up is too small for even the smallest child's fingers (B).

Frameless glass doors have metal rails or patches carrying the pivot or pivot housing. The rails and patches are thicker than the glass, so even if the pivot is fitted as close to the edge as is compatible with the operation of the door, a gap must occur by the glass as the door opens. If a concealed door closer is used in the head of the door the pivot will be about 60 mm ($2^3/8$ in.) from the edge of the door, unless a portion of the closer can be lost in the mullion. Aluminium doors require cleats to join the stiles with the rails. These cleats prevent the pivot being as close to the heel as with wooden doors.

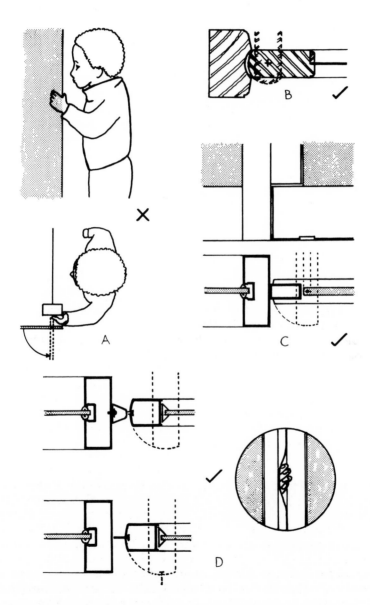

Figure 12.8 (A) Shopfronts may hold this danger; (B) safe operation of wood swing door; (C) eliminating the gap on a glass door; (D) use of finger guards on aluminium doors.

To prevent crushing of fingers, either the gap that occurs with glass and aluminium doors must be eliminated in some way or a resilient material that will deform under pressure must be fitted to the edge of the door or to the frame. A method of eliminating the gap with glass doors is to project the top and bottom rails beyond the edge of the glass and fill the space between the door and frame with either an aluminium box section (as shown at C) or by glass (as in an all-glass shopfront). Alternative methods of providing a strip that crushes more readily than fingers are shown used with aluminium doors (D). Both types of strip (or finger guards) are made of neoprene or other rubbery material.

Heavy swing doors with strong closers can harm infirm, frail and handicapped persons (fig. 12.9). Electro-mechanically operated door springs actuated by push button controls are available. The door opens automatically and is held open long enough for an elderly or handicapped person to pass slowly through. Other persons can open the door manually, it then closes automatically.

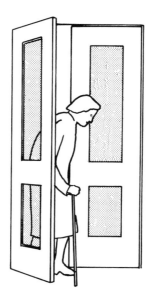

Figure 12.9 A pair of doors, framed in the solid, of the design shown are to be found in a home for old people. Hollow core construction would have given a lighter door with virtually the same appearance.

12.4   Siting doorways

To avoid struck-by accidents, hinged and pivoted doors must not swing into space that is used solely for traffic. Designers must also consider possible unplanned use of space by occupants. For example, any office is likely to need filing cabinets, and thought must be given to where they are going to be placed (fig. 12.10).

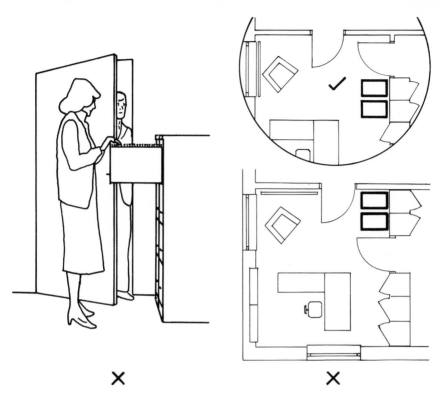

Figure 12.10 With the drawer of the filing cabinet open a blow from the door could topple the cabinet and cause serious injury. The insert shows a way of avoiding the danger.

If persons have to stoop or squat to use a fitment in, or close to, a space into which a door swings, the risk of an accident is increased. The potential danger of a cooker near a doorway, as in fig. 12.11, should reveal itself at the design stage. Yet such situations are not uncommon, perhaps because the swing of a cooker door is not drawn on kitchen plans. Cupboards are usually shown on plans with their doors partially open, perhaps this is why they do not draw attention to the possible danger they present (fig. 12.12).

Doorways close together on adjoining walls are awkward to use and hence not good ergonomically (fig. 12.13). Fire escape doors must open outwards but they should be recessed so that they do not project into traffic circulation or impede escape routes (fig. 12.14). Siting doorways so that they give direct access to roadways can lead to accidents due to momentary inattention on the part of users (fig. 12.15). With up-and-over doors the principal danger to avoid is that the edge of the door not fully open, and thus above eye level, may not be seen (fig. 12.16).

Figure 12.11 When cookers are involved in struck-by accidents there is a risk of a secondary accident.

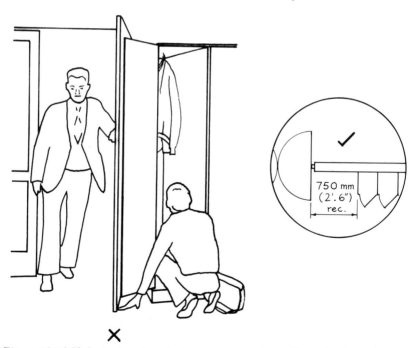

✗

Figure 12.12 If the locker door had been hung on the other side the problem of interacting doors would not have arisen. It would not however have removed the risk of injury to the student's hand. Ample clearance should be provided.

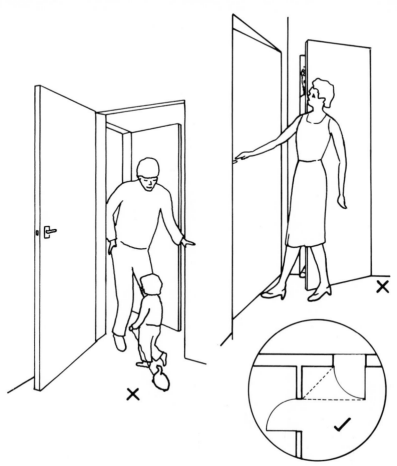

Figure 12.13 People get in each other's way when doors are close together and struck-by accidents are likely to happen. A recommended minimum spacing is shown.

Where doorways are close to the head of the stairway, users cannot be adequately cued to prevent them coming upon the stairway unexpectedly. The need for warnings and cues has been explained in section 2.4 and ways of making stairways prominent are given in chapter 11. Where the rise of flight does not exceed 600 mm (23½ in.), the consequences of a fall on the steps are not likely to be too serious, except perhaps for the elderly, so less warning of the presence of the steps might be acceptable than if the fall were greater. In such circumstances a landing 625 mm (24½ in.) deep beyond the doorway is needed (fig. 12.17). This depth equals one average pace of adult males. A young man bursting through the doorway should have time to react, an elderly woman will be moving more slowly and will be able to steady herself before starting the descent.

Figure 12.14 Projecting doors impede escape routes and cause struck-by accidents.

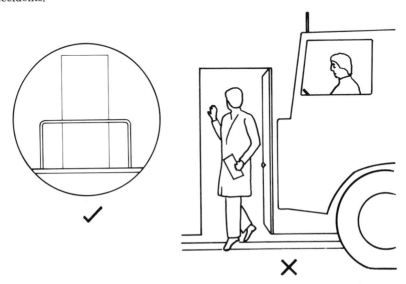

Figure 12.15 Keep doorways away from service roads or provide barriers to cause persons to make a turn before stepping into the roadway.

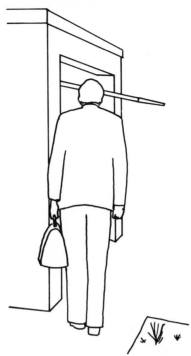

Figure 12.16 If the edge of the up-and-over door is between eye level and the top of the head of a person approaching he may walk into it. Paths should not direct walkers into the space in which the door moves.

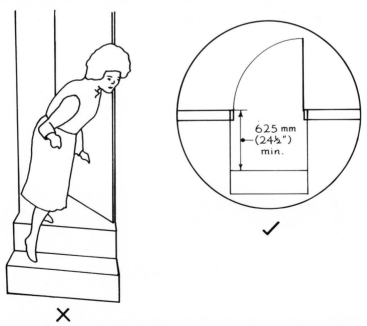

625 mm
(24½")
min.

Figure 12.17 At an external door of a house a small difference in levels (under 600 mm (23 in.)) is likely to be anticipated; indoors it can be a booby trap for the unwary so a landing must be provided.

Where the rise is greater than 600 mm (23½ in.) no part of a doorway or the area in which the door swings should be nearer than 800 mm (2 ft 7½ in.) to the top or bottom of a domestic stairway, or nearer than 1000 mm (3 ft 3½ in.) if the stairway is used by many people, say over fifty. The need to keep doorways away from the bottom of a flight is emphasised by the possibility of several accidents, as shown in fig. 12.18. A danger that can arise at the head of a flight from the proximity of a doorway is shown in fig. 12.19.

When hinged doors are used above work areas, sooner or later they will be left open for someone to bump into, as shown in fig. 12.20.

12.5 Door furniture

Lever handles on doors (fig. 12.21) are fixed at heights above floor level varying between 900 mm (2ft 11½ in.) and 1220 mm (4 ft). The average elbow height of a woman wearing shoes is 1000 mm (3 ft 3½ in.), so handles at around this height are likely to catch in sleeves at the wrist. If fixed higher, they are at a height to catch short sleeves. At around 900 mm they are likely to get caught up

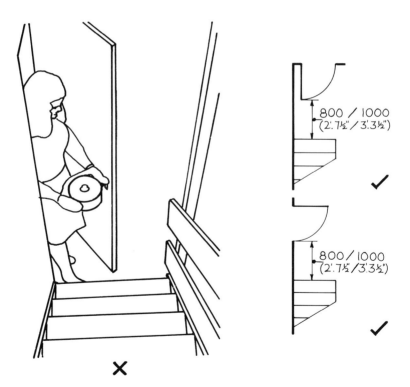

Figure 12.18 Not only might an accident be caused by someone coming through the doorway at the moment that a person is descending the stairway (an exuberant young person, perhaps), but if an accident occurred on the stairway the victim might be injured by falling against the door.

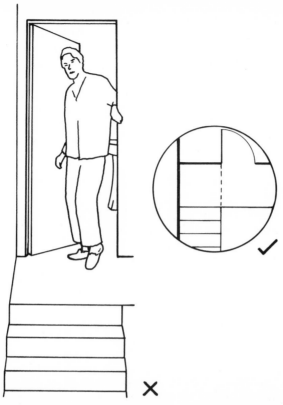

Figure 12.19 A doorway opposite the head of a stairway is best avoided especially if it is a bathroom doorway: someone feeling unwell in the night may visit the bathroom and be overcome by faintness on leaving.

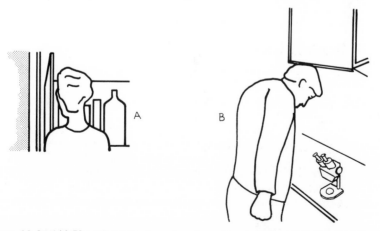

Figure 12.20 (A) If cupboards above work areas are at the same level of the user's face the edge of the door might not be readily distinguishable against other verticals. Existing cupboard doors should have their edges painted in a bright, contrasting colour; (B) when a person strikes his head against an open door the impact can 'knock him silly', he can also suffer a badly cut head; the corner of the door is very dangerous.

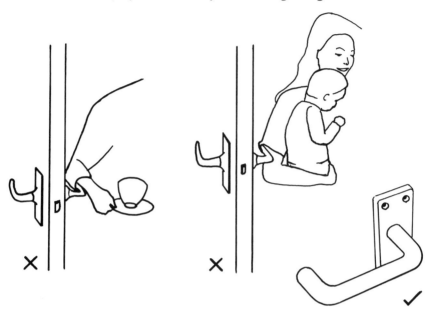

Figure 12.21 The dangers of lever handles, and the solution.

in other parts of loose clothing, to the detriment of the clothing even when no injury results. The solution is to use a handle that returns to within about 3 mm ($^1/_8$ in.) of the face of the door.

With knob handles, no part should be nearer than 75 mm from the edge of the door, to avoid the barking of users' knuckles on pulling-to the door behind them.

SECURITY

12.6 Attacks on doorways

For the security of a doorway, the door, frame and locking devices are mutually dependent. Locks and fasteners are described in the next chapter, here they are considered only so far as they affect the construction of doors and door frames.

Being an obvious way into a building, illegal entry is common via a doorway. Considerable damage may be done to the door in the process of forcing the lock, using methods described in the next chapter. Doors also suffer damage through being deliberately punched and kicked, and from attacks on door furniture. Additionally, they are forced open by occupants who have an irresponsible attitude to the building they are using and have forgotten, mislaid or lost keys — damage to doors in sleeping rooms was found to be the most pervasive and costly type of damage in U.S. naval bachelor enlisted quarters (Stroik, J. *ed.* (1981) *Building Security.* Philadelphia: American Society for Testing and Materials, Brady. C. *et al.* 'Reducing vandalism in naval bachelor enlisted quarters', 49—59).

The interdependence of door, frame, lock and hinges is recognised in the U.S. standard ASTM F476—76, *Standard Test Methods for Swinging Door Assemblies*, which defines methods of testing the capability of a complete, swinging (side-hung) door assembly to restrain or frustrate a break-in by opportunist or semi-skilled burglars. There are four grades of resistance, the lowest that of a door assembly adequate for single-family residential buildings located in stable, low crime areas; the highest a door assembly suitable for small commercial buildings located in high crime rate areas, and also suitable for residential buildings having an exceptionally high incidence of semi-skilled burglary attacks. The grades between provide, in the words of the standard, low-medium and high-medium security. The tests are not intended to provide a measure of resistance against skilled burglars attacking high pay-off targets.

12.7 Siting doorways

The need to locate entrance and exit doors, as far as possible, where they are overlooked has been emphasised in section 8.8. Front doors to houses will generally be visible to neighbours and passers-by, but there is a danger in providing storm doors to an unlocked vestibule, as shown in fig. 12.22. Back doors tend to be more secluded than front doors, and hence more vulnerable to housebreaking. Doors in entries, passageways and garages, where the burglar is hidden from view and can take his time about finding a way to open the door, should be avoided if at all possible.

'Thinking like a thief and vandal' should lead the designer to

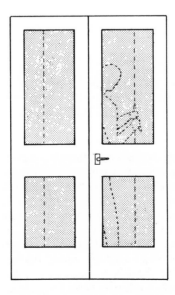

Figure 12.22 Storm doors may give cover and mask the noise of a break-in.

consider the relationship of a doorway to other features of a building. Illegal entry *through* the door is not the only factor to guard against. As shown in fig. 12.23, entry can be by means *of* a door. In this case the balcony assists the thief, but an agile man can get up on to a flat roof in a similar way. This is not the only reason why door pulls or handles should not be at knee height on up-and-over doors — in this position their accessibility makes them vulnerable to vandalism.

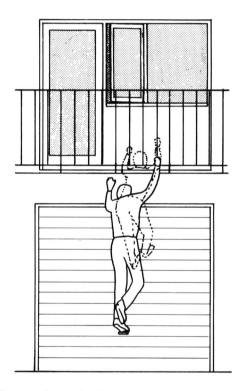

Figure 12.23 Thieves and vandals like to be able to get a foot on door furniture.

12.8   Door construction

Install easily penetrated glazing in a door or nearby, or fix a letter box or cat door through which an arm or implement can be inserted, and you will probably nullify the protection of a stoutly constructed door fitted with a good lock. If the door is glazed with beads on the outside, or if outside putty is still fresh, it may not be necessary for a burglar to break the glass, he will be able to remove a whole pane (see fig. 6.2). In panelled doors the panels may not offer much more resistance to breaking or being removable than glass. In flush doors a hardboard facing can be broken with blows from a hammer, kicked in, or penetrated with a sharp tool and nibbled away with a pair of pincers, or even pulled away by hand. If the door is of light construction having a core of lattice or honeycomb framing (the so-called 'hollow core') the whole door is easily destroyed. Boarded doors can be taken apart by prising off the boards. Vandals may be bent on the complete destruction of a door, burglars will be content with breaking free the lock. Cupboard doors of veneered chipboard are easily vandalised by being pulled off their hinges because of the poor holding power of screws in chipboard.

For entrance doors, any solid core door (fig. 12.24) is much stronger than a hollow core door, because of the support given to the facing material. A solid core of wood is more resistant to attack than a core of chipboard, a plywood core is the most resistant wood construction of all and gives strong security. For extra protection, a wood door may be faced with a steel sheet, 1.5 mm ($^1/_{16}$ in.) is recommended, or a steel sheet may be incorporated in the core. Steel edge strips protect the bolts of locking devices from being broken out of the door or frame.

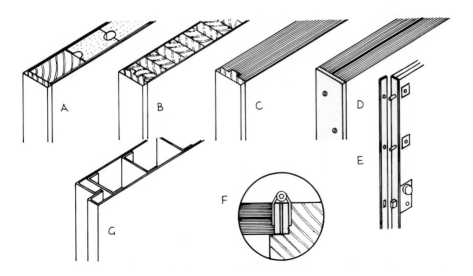

Figure 12.24 Vandal and burglar resistant doors. (A) Extruded chipboard core; (B) laminboard core; (C) plywood core; (D) plywood core with hardened steel plate and steel edge strip; (E) edge strips holed for bolts and lock; (F) hinge detail; (G) steel door.

Steel security doors of stout construction, as in the example shown in G, fig. 12.24, provide strong security. A rebated edge that lips over the door frame gives additional protection to the bolt of the lock of an outward opening door.

Where bullet resistance is required of a door, the core may incorporate one of the bullet-resistant materials described in chapter 8, or a steel door may be bullet resistant if faced with heavy steel. The 'Picus' bullet-resistant door, which has a core of two 18 mm

(¾ in.) thick sheets of plywood with a 3 mm (⅛ in.) thick steel sheet bonded between them, passes the classes G0, G1 and S of BS 5051 (see sections 6.3 and 8.11).

12.9 Door frames

Door frames need to be of stout wood or metal construction. Light hardwoods, softwoods and aluminium alloy frames will give basic security, provided wood sections are ex 100 x 75 mm (4 x 3 in.) and aluminium frames have a wall thickness of at least 2.5 mm in places where the metal is likely to be attacked to force the lock or hinges; alternatively other precautions can be taken with aluminium doors (see section 13.7). For strong to maximum security, medium to high density hardwoods and steel should be used. The jambs of pressed steel frames should be filled with fine concrete, as shown in A, fig. 12.25. Rebates should be at least 18 mm (¾ in.) deep, and worked from the solid in wood frames.

The fixings of frames should not be more than 450 mm (18 in.) apart. Frames built-in as walls are constructed should be secured by stout anchors. If, on wood frames, anchors are fixed into a groove in the back of the frame, as shown in A, fig. 12.25, they will hold the frame secure even when the screws do not have a good hold in the wood — bricklayers seldom carry screwdrivers in their tool kits so when the bricklayer himself fixes the anchors to suit his courses of brickwork he is likely to drive in the screws with a hammer. Dowels should be used at the foot of the frame and the horns of the head built in at the top.

When frames are fixed after the walls have been constructed, as is the case with polished hardwood frames, for strong security, and better, the initial fixing by screws should be supplemented by steel

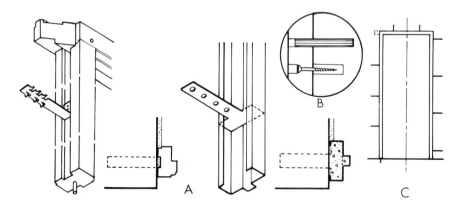

Figure 12.25 (A) Anchors for wood and metal door frames; (B) post-construction fixing supplemented by steel rods driven through wood frame; (C) position of anchors and other fixings.

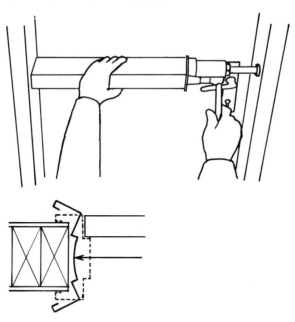

Figure 12.26 Hydraulic car jack and suitably sized piece of wood used to spread joints of door frame.

rods of 10 mm ($^3/_8$ in.) diameter driven into holes drilled through the frame into the wall. The common practice of fixing frames by nailing into wood pads built into brickwork is not to be relied on, even for basic security, the pads shrink and are unlikely to withstand battering of the frames or resist them being prised away from the fixings.

Hydraulic and screw jacks can be used to distort or force apart the jambs of a door frame sufficiently to free the bolt of the door lock. The sectional view in fig. 12.26, based on an illustration by Martin and Patrick (Stroik, J. *ed.* (1981) *Building Security*. Philadelphia: American Society for Testing and Materials. Martin A. and Patrick, C. 'Jamb/stud wall security' 177—189), shows how a metal door frame can be deformed by pressure from a jack. Frames in a timber framed wall, as in a house where this construction is used, can have their jambs, wood or metal, spread apart. Frames in a masonry wall not backed up tightly may also spread sufficiently to free the bolt. Tests by Martin and Patrick have shown that with the usual timber framed construction of studs with plywood one side and plasterboard the other, under pressure from a jack in a doorway, the plywood and plasterboard pull away from the studs and, as pressure is increased, the studs rotate due to the unequal strength of the plywood and the plasterboard. This rotation can cause failure of the lock by breaking the bolt. Horizontal timbers (nogging pieces) fixed between the studs at lock height increase resistance to spreading and rotation, but when a doorway is near

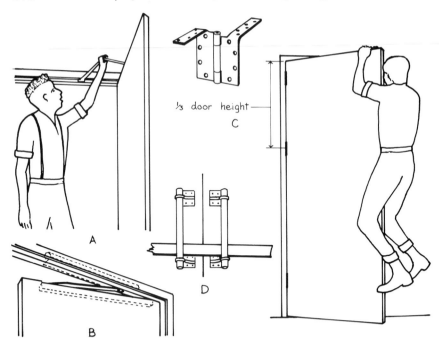

Figure 12.27 (A) Door closer arms are a temptation to the vandal; (B) concealed closer (Relcross) will be little noticed even when door is open; (C) put middle hinge above centre for greater strength or use anchor hinge (Hager) at top; (D) avoid bow handles where this sort of fooling about is likely.

the end of a wall there will be few studs to resist the pressure transmitted through the nogging pieces.

12.10 Door furniture and fittings

Items fixed to doors are targets for vandals rather than burglars, and, unlike burglars who attack doors only when they are closed, vandals attack both closed and open doors. There is more scope for destruction when a door is open, the panels or facing can still be kicked in, and with the vandal inside the building he has access to door closers, as shown in fig. 12.27. By swinging on the door the vandal may pull it away from its hinges. Three hinges are necessary for basic security and raising the middle hinge from its customary centre position to approximately the position shown in C will give strength where it is most needed. Lever handles and bow handles lend themselves to being levered off; another piece of nonsense is to block the entrance or exit of people through swing doors, as shown in D. Stout knob handles securely fixed offer the best vandal resistance.

Chapter 13

Locks and Fasteners

13.1 Symbolic and real protection

Where a lock is intended merely to give notice that entry is not desired, it may be of a most simple type. Where greater security is needed but unauthorised entry by the use of force is unlikely, locks that will operate only with the proper keys are required. Where force is likely, locks must be strong enough to resist hammer blows, leverage, drilling and hacksawing. The picking of locks that are reasonably proof against the use of false keys is uncommon, outside the use of skeleton keys. Films and TV give a most misleading impression of the ease of opening locks with a probe or piece of bent wire, real villains usually prefer to use force.

The selection of a lock for a location where force might be used cannot be divorced from the design of the door, the frame, and its fixing. For the moment we are concerned only with locks and fastenings, so cross-referencing is used to draw attention to the interdependence of locks and the items they are fixed to. Special locks for windows are dealt with in the section on windows in chapter 14.

13.2 Definitions (see fig. 13.1)

Rim lock: a lock screwed to the face of a door.
Mortice lock: a lock housed in the thickness of a door.
Dead bolt: a lock bolt with a square end, it cannot be pushed back once it has been moved out by the key, only the key will bring it back without the use of force.
Spring bolt or latch bolt: a lock bolt bevelled at the end, it normally protrudes from the body of the lock. It can be pushed back and will return without turning a key or handle.
Claw bolt: an expanding bolt used in locks for sliding doors, and on pairs of doors to prevent them being pushed open when locked but not bolted.
Hook bolt: another bolt used for sliding doors.
Dead lock: a lock with a dead bolt only.
Sash lock: a mortice lock with handle and key on the same vertical line.

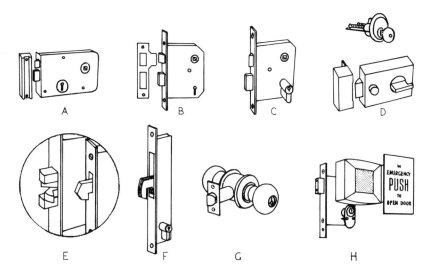

Figure 13.1 (A) Rim lock and staple; (B) mortice lock and striking plate; (C) cylinder mortice lock; (D) cylinder rim night latch; (E) claw bolt and hook bolt; (F) lock for narrow stile, metal sliding door; (G) knob lockset (key-in-knob); (H) emergency exit lock and alarm (Kaba).

Latch: a kind of lock having only a spring bolt operated by a handle.
Night latch: a lock with a spring bolt withdrawn by means of a key outside and a handle inside.
Cylinder lock: a lock with the mechanism in a cylinder. Hence *cylinder rim night latch*.
Knob lock set (key-in-knob): locking knobs operated by a key or by buttons that are pushed or turned, or both pushed and turned.
Emergency exit lock and alarm: a key operated lock that can be opened in an emergency by pushing a pad or bar, causing an alarm to sound. Pushing the pad of the lock shown in fig. 13.1 also breaks the glass cover on the alarm switch.

13.3   Mechanisms

Cheap locks have mechanisms such as shown in A, fig. 13.2. The tumbler prevents the bolt being pushed back once the key has been turned. A double-hand tumbler, shown in B, used with a double-ended keyhole, enables a rim lock to be fixed either right or left-handed. Both parts of the tumbler lift, whichever part the key acts upon. The notches in the keys are merely for appearance. There is no 'differs', all locks of one make can be opened with the same key. Any other key of a suitable size will also open the lock. A thief who does not have a key can pick the lock by lifting the tumbler with a piece of bent wire and sliding the bolt back.

Wards
To make the use of keys other than the proper one more difficult,

wards may be used. These are obstructions that the key must clear as it turns (fig. 13.3). Key holders can easily make their key into a skeleton key that will pass other locks in the same series, even though the wards are different. An accomplished burglar who carries a selection of skeleton keys hoping to have one of a suitable size to open any warded locks he meets may be foiled by a shaped keyhole because of the distinctive shape of key needed (B, fig. 13.3), unless he is expert enough to carry very thin keys for this reason.

Levers

Lever mechanisms are distinguished from simple tumbler mechanisms in that levers have to be lifted a precise amount for the lock to work, whereas the overlifting of a tumbler does not matter. Figure 13.4 shows a mechanism consisting of three levers and a dead bolt. As the key turns it lifts the levers and at the same time moves the bolt. The bolt stump passes from one pocket in the levers to another through a 'gate'. With the stump trapped in a pocket the bolt cannot be locked or unlocked without the key. The lock is intended for use from both sides, thus the notches in the blade of the key are symmetrical; whichever side the key is inserted, one half of the blade operates the levers while the highest step on the other half operates the bolt. By having a number of levers, each having to be lifted differently, protection against the use of false keys is obtained. In manufacture, the levers can be assembled in different combinations and different levers can be substituted, thus producing a large number of differs.

However, the degree of security given by a three lever lock is not high; BS 3621: *Thief Resistant Locks*, requires at least five levers. If there are less than eight levers, some device to increase the resistance

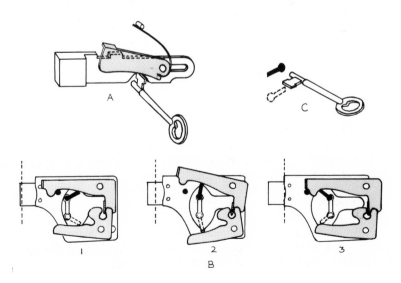

Figure 13.2 (A) Tumbler mechanism; (B) double-hand tumbler, key may be used whichever way it is inserted; (C) sideways keyhole for double-hand use.

Figure 13.3 (A) Wards, key notched to pass, and how a skeleton key passes; (B) corrugated key and keyhole.

to picking or manipulation by means other than the correct key must be included. False notching of levers so that the bolt stump catches in the notches if the levers are not lifted sufficiently is a way of increasing resistance (fig. 13.5). BS 3621 also requires a minimum of 1000 differs. These must be effective differs; tolerance in the mechanism, wear of levers and of keys must be allowed for to avoid the possibility of a key operating a lock for which it was not intended (see also section 13.8).

A different lever mechanism from the 'English' type, which has two pockets for the bolt stump, is also shown in fig. 13.5. The bolt stump moves to the end of the levers, its forward movement being arrested by a step on the bolt meeting the case. When the bolt is locked, the bolt, bolt stump and levers form a continuous line of metal, thus offering high resistance to the forcing back of the bolt by jemmying. The boxed striking plate used with the lock gives

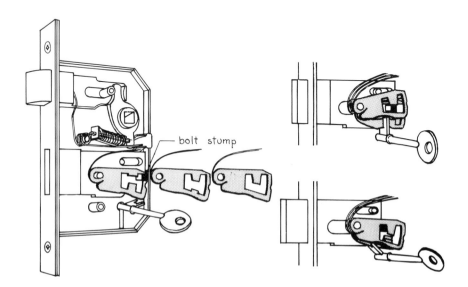

Figure 13.4 Movement of bolt of mortice lock with three levers.

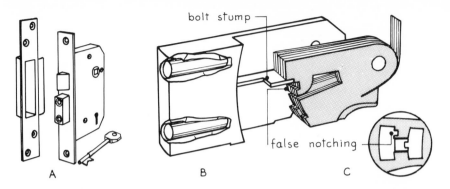

Figure 13.5 (A) Lock with boxed striking plate; (B) deadlock with hardened steel rollers to prevent hacksawing and levers with false notching as an anti-pick device (ERA); (C) false notching on English type lever.

further protection against this sort of attack, and the hardened steel rollers in the bolt provide the resistance to five minutes of hacksawing required by BS 3621.

Pin tumblers
In the mechanism of a cylinder lock (fig. 13.6), pin tumblers perform a similar function to the levers of a lever lock, and like levers they have to be lifted a precise amount for the lock to operate. Each pin tumbler is in two parts: the driver or roller at the top and the pin at the bottom. While the drivers are held down by springs they prevent the inner cylinder or plug from turning, insertion of the key into the plug raises the pin tumblers so that all the joints between the drivers and pins coincide with the joint between the plug and the outer cylinder. The plug can then be rotated, it takes the pins round with it and turns the connecting bar that operates the bolt in the lock case. The two shaped drivers shown in the figure are an anti-pick device, they tilt and jam if not properly aligned when an attempt is made to rotate the plug.

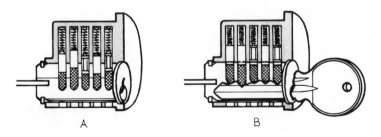

Figure 13.6 Cylinder lock (Yale). (A) Springs holding pin tumblers down to lock inner cylinder; (B) key pushing pin tumblers into unlocked position.

As many as 24 000 differs are obtainable as standard with cylinder locks. Thus the familiar cylinder rim night latch is well protected against false keys. It can, however, be picked by an expert and is very vulnerable in two other ways. The spring bolt is easily pushed back by end pressure and the inside handle can be operated if it can be reached by breaking a pane of glass or making a hole in the door. The stop knob or snib provided to hold the bolt retracted or projecting allows the bolt to be deadlocked, but only from inside, it also prevents the use of the key from outside. Another means of deadlocking with release from the outside is required for at least basic security. In table 13.1 the different systems of deadlocking and of preventing a thief from unlocking the door by turning the handle are listed. With automatic deadlocking, an auxiliary bolt is sprung back on striking the staple as the door is closed and this operates the deadlocking mechanism on the main bolt, also a lock fitted into the handle enables the handle to be locked from inside with the same key as used outside. With the so-called semi-automatic deadlocking, the bolt is deadlocked every time the door is closed, but to lock the handle the key has to be inserted from outside and given an opposite turn. With non-automatic deadlocking, the bolt and handle are deadlocked simultaneously from outside, either by giving the key an opposite turn or by turning it twice.

Because a lock has a deadlocking facility this does not necessarily mean that it has adequate security in this respect. To meet the requirements of BS 3621 a 'thief resistant' lock must remain secure after a force of 9000 N (2023 lbf) has been applied for one minute to the end of the bolt in the direction in which the bolt moves to unlock. The lock need not be capable of further operation after resisting this force.

Kaba locks
Several cylinder locks with more complicated mechanisms are available. Some examples are to be found in the Kaba range. Kaba cylinders are a Swiss invention, they have a mechanism developed from that of the ordinary cylinder lock. One version of the Kaba mechanism is shown in fig. 13.7, this has three rows of pin tumblers but Kaba cylinders are also available with two rows (one each side of the key) and with four rows (two each side, one at 90° and one at 45° to the key face). With up to five pin tumblers of four different legnths to a row, billions (U.S.) of differs are possible. Master-keying is readily accomplished, it is possible to make a master key that will fit 10 000 cylinders, each operated by a different key. Picking resistance is high, mainly because of the number of pin tumblers employed but also because of other features that reduce the possibility of picking. For instance, all pin tumblers have equal spring load when probed from the keyway, and specially shaped anti-pick pins are introduced — one is shown in the sectional view in fig. 13.7.

Table 13.1 Deadlocking of spring bolts and handles of cylinder rim night latches

Action	Deadlocking	Release of deadlocking	Locking of handle	Remarks
Automatic	Automatic when door closes	By key outside and handle inside	Separate lock inside	Handle lock does not interfere with operation of main lock so keyholder can lock handle on leaving
Semi-automatic	Automatic when door closes	By key outside and handle inside (if handle is not locked)	By key from outside	Handle cannot be locked or unlocked from inside
Non-automatic	By key from outside	By key from outside	By key from outside	Positive action necessary to deadlock bolt Handle cannot be locked or unlocked from inside No deadlocking from inside (except by snib)

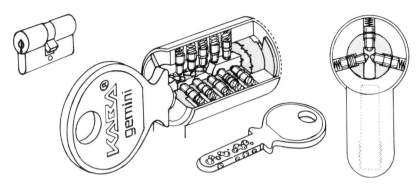

Figure 13.7 Kaba Gemini cylinder for use with a mortice lock.

Abloy lock

The Abloy cylinder shown in fig. 13.8 has a form of disc mechanism derived from a combination lock. It is of Finnish origin. The lock is cheap to make and assemble, there are no springs and the mechanism is insensitive to moisture, frost and impurities. Over 360 million differs are available to the standard key of semi-circular cross-section. Master-keying is possible over a most extensive range. Moreover, the makers say that picking is too difficult for it to be accomplished. The cylinder is fitted into lock cases of many different types: rim, mortice, pad and others.

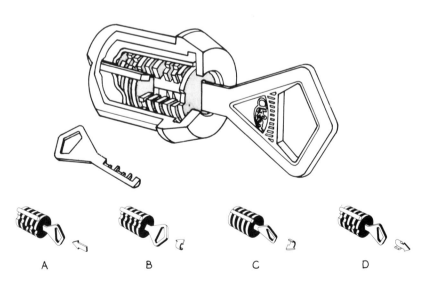

Figure 13.8 Abloy cylinder. (A) Key inserted; (B) turn brings discs to the correct combination; (C) locking bar pressed into slot of cylinder drum which can then rotate and open lock; (D) key turned in opposite direction brings cylinder parts back to original position.

Knob lock sets (key-in-knob)

The security weakness of this type of lock is that the locking mechanism is contained in the handle, not within the door. Entry may be obtained by breaking off or sawing off the handle. These locks are popular for hotel bedrooms, where perhaps banging and sawing might attract attention. They have the advantages of being available with a variety of locking functions (fig. 13.9) and being easy to install: their installation is carried out mainly by boring, aided by jigs and strike locators.

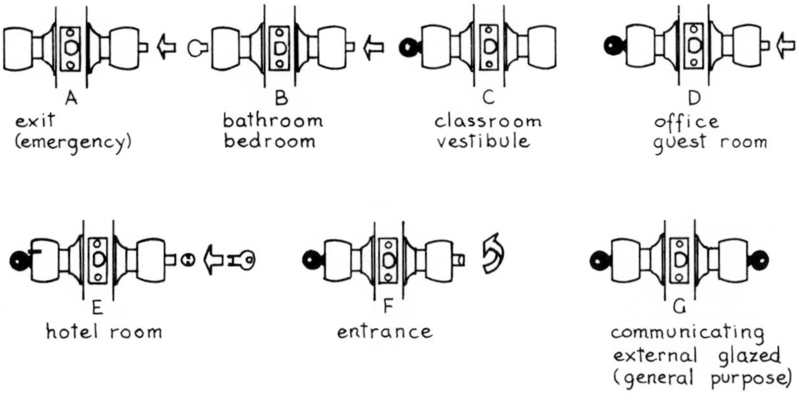

A
exit
(emergency)

B
bathroom
bedroom

C
classroom
vestibule

D
office
guest room

E
hotel room

F
entrance

G
communicating
external glazed
(general purpose)

Figure 13.9 Knob locksets (key-in-knob) and doors they are used on. (A) Push button locking with prevention of lock out; (B) privacy lock, can be opened by emergency turnkey or screwdriver; (C) inner knob always free to turn; (D) as A but locked or unlocked by key from outside; (E) entrance by key only. When locked from inside by push button, indicator pin prevents insertion of keys other than emergency key. Rotation of push button by management prevents use of room key by guest. (F) Push button and key locking as at D. Pushing and turning button requires use of key for unlocking until button manually restored to unlock position. (G) Arranged for dead locking, by one or both keys, in various ways.

13.4 Multi-point protection

For strong security, one bolting point on a door is not sufficient. A burglar can attack a door from the hinge side, ignoring the lock, and on an outward opening door either by removing the pins of the hinges or sawing through the hinges to gain entry. Hinge, or dog bolts (fig. 13.10) give protection against this form of attack, they come into action automatically when the door is closed. Added protection can be provided on the opening edge by surface bolts or by the less obtrusive mortice bolts of the type shown in D in the figure. For final exit doors it is essential to have bolts that are lockable, for convenience preferably operated by the same key as the main lock. Pinion key operation is suitable for doors locked from the inside. The bolts may be fitted on the top and bottom edge of the

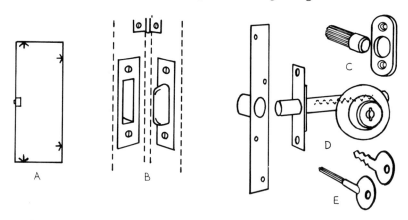

Figure 13.10 Multi-point protection. (A) Bolting points on door; (B) heavy duty hinge bolt (Ingersoll); (C) hinge bolt (SS Products); (D) lockable mortice rack bolt; (E) pinion key for a non-lockable mortice rack bolt.

door instead of on the opening edge; as their position will not be apparent to a burglar from the outside it is best to avoid putting them in obvious places.

Multi-throw locks (fig. 13.11) are more convenient for the locker-up because he has to turn only one key, unfortunately they

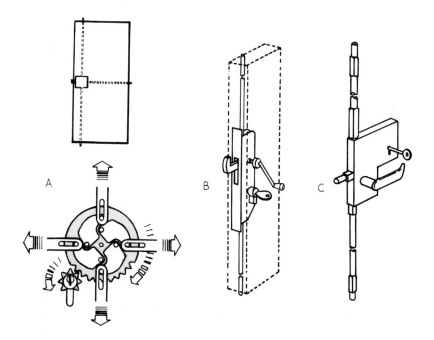

Figure 13.11 Multi-throw locks. (A) Mul-t-lock mechanism; (B) sliding patio door lock (Schlegel); (C) lock for wooden door (Surelock).

are also more convenient for the burglar because he has only one lock to defeat. If the latter uses force and pushes back one of the bolts it may bring the other bolts with it. The mechanism shown in A is designed to prevent this happening. Top and/or bottom bolting is particularly necessary for sliding doors, to avoid relying entirely on the holding power of a hook or claw bolt: sliding doors are easier to lever open than side-hung doors.

Double doors always require top and bottom bolting on the first closing leaf. On glazed double doors, flush bolts should be fitted into the edge of the meeting stile (fig. 13.12) so that the bolts are concealed when the doors are closed. This is also the preferred location for bolts when the doors are not glazed because a more secure fixing is obtained. Wherever practicable, meeting stiles should be rebated so that the dead bolt of the lock is covered. However, rebating is a complication when both leaves are in use, and a door selector (co-ordinator) may be necessary to ensure that they close in the correct sequence.

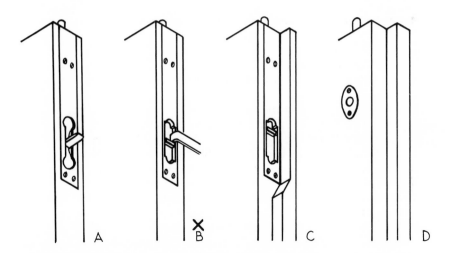

Figure 13.12 Bolts for first-closing leaf of double doors. (A) Lever action, advantageous for non-rebated doors; (B) slide bolt, release from outside is possible; (C) fixing complicated by rebate; (D) mortice bolt, burglar needs key, so does occupier.

Up-and-over garage doors should have a bolt each side for fastening from inside. If used as a final exit door, a good quality lock each side is advisable, some of the locks fitted to stock doors give little security.

13.5 Panic bolts and panic latches

These are devices designed to operate under the application of simple forward or downward pressure at any point on a bar that extends across the doorway (fig. 13.13). They are a security weakness

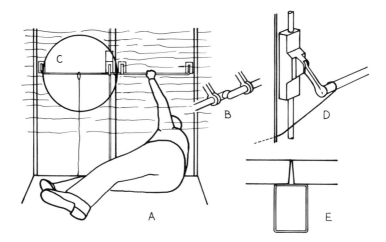

Figure 13.13 (A) Double panic bolt designed for release from below smoke in a fire emergency; (B) bevelled joints in push bar of double panic bolt; (C) string left in a position for an intruder to use; (D) single panic bolt released by wire through butt joint of double doors; (E) removable mullion.

because of the way that the bar can be operated by a piece of string or wire. When string is used it will have been previously positioned from inside — which may not be difficult in a building used by the public. It can be secreted by the bolt, rather than left in the prominent position shown in C, with the end poked under the door.

On a single door the push-bar may operate a latch. On double doors with rebated meeting stiles a double panic bolt such as that shown in A must be used. There is a push-bar on each door but a bolt on one door only. Between the two push-bars is a bevelled joint as shown in C. Operation of the left-hand push-bar opens the left-hand door, operation of the right-hand push-bar opens both doors. If the doors are automatically closed then a door selector (co-ordinator) must be fitted to ensure that the doors close in the correct sequence. The door selector is vulnerable to vandalism, for this reason and because of the high probability of malfunction of the double panic bolt, a butt joint between the meeting stiles of the two doors is usually preferred, single panic bolts can then be used. Security is then in greater jeopardy because the push-bar can be operated without prior arrangements or risk of detection by a piece of bent wire pushed through the joint, as shown in D. This is not possible if the meeting stiles are rebated.

Requirements for the performance and testing of panic bolts and panic latches mechanically operated by a horizontal push-bar are given in BS 5725, Part 1 (1981) *Emergency Exit Devices*.

Doorways used for deliveries, to a bar for example, often also function as emergency exit doors. If these are double doors fitted with panic bolts they may not only provide an easy way in for the burglar but lead him straight to his plunder. A mullion between double doors will allow single panic bolts or latches to be fitted, without leaving a gap between the meeting stiles through which a wire can be inserted. A removable mullion (E, fig. 13.13) can be taken away when more than one door is needed for access. In an emergency the mullion can speed rather than hinder escape. Without it a pair of doors can have a funnelling effect pushing people to the centre of the opening, with two single doors this does not happen. However the fire prevention officer will have to be convinced of this.

13.6 Padlocks

The advantage of padlocks is that they can be used with hasps and staples that are easily installed, or with chains that require no installation whatever. This makes them popular as an improvised and cheap fastener, but they have the disadvantage of being detachable and portable. Users are liable to carry them off accidentally and potential thieves can substitute a lock, for which they have a key, for the original. Also padlocks are easier to force than fixed locks, and some of the fittings they are used with are ridiculously easy to circumvent. One situation in a planned building where the use of a padlock is justified is where the user has to provide his own lock because of frequent changes of user, another situation is where a fixed lock would be difficult to install, on a roller shutter for example.

The shackle of a padlock may slide in and out of the body or it may be hinged. The body may be constructed of iron, steel in various forms — sheet, laminated, cast, drop-forged — brass or bronze. Shackles of good quality locks are of hardened steel or manganese bronze. Close shackle and concealed shackle padlocks resist cutting and prising open better than open shackle padlocks (fig. 13.14).

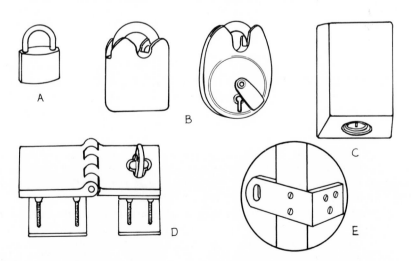

Figure 13.14 Padlocks. (A) Open shackle; (B) closed shackles (Yale and Chubb); (C) concealed shackle (Chubb); (D) locking bar with back plate and concealed bolts (Chubb); (E) one of a pair of locking plates with concealed fixing.

The locking mechanism may be based on wards, levers, discs or pin tumblers.

Padlocks will not give maximum security. For strong security a concealed shackle padlock with a strong locking bar of the type shown in D should be used; the lock mechanism should have about ten levers or the equivalent. For basic security padlocks must be close shackled with a strong body and at least a five-lever, or equivalent, mechanism. The locking bar must not have fixings that can be removed from the outside of the door, and it must be resistant to prying off.

13.7   Attacks on locks and fasteners

Picking and forcing

The basic technique of lockpicking is to keep the lock mechanism under light pressure while manipulating levers or pin tumblers with a pick. Slight irregularities of manufacture and clearances between working parts enable slight movements of the mechanism to be detected, indicating that individual levers or pin tumblers are in the release position. Wear in the mechamism increases the lock's vulnerability; worn cylinder locks can be opened by putting the plug under slight pressure and 'bouncing' the pin tumblers into position by rapidly drawing a pick along the keyway. But lockpicking requires patiently acquired skills and time to work undisturbed on the lock, most burglars prefer to use force.

Some of the implements used by burglars are shown in fig. 13.15. Occupiers often provide what is necessary themselves by leaving garden tools in unsecure sheds and outbuildings. Hook bolts on sliding doors can often be released by levering up the door at the

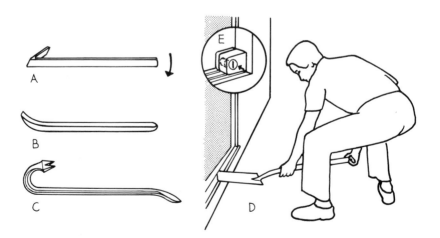

Figure 13.15 Tools used to overcome locks by force. (A) Pry bar used between lock and striking plate; (B) and (C) crow bar and case opener (jemmys); (D) use of occupier's own tools; (E) additional lock for patio door (Chubb).

bottom with a spade; additional locks on patio doors may be necessary to give protection against both lifting and sliding by forcing with garden tools or the other implements shown.

Rim locks

Exposed screws on rim locks are vulnerable to tampering, either as an act of vandalism or to assist subsequent entry — the hold of the screws can be weakened so that the door is easily forced. Back plates and provision for screwing into the edge of the door, as shown in fig. 13.16, are answers to this threat.

Figure 13.16 Tampering with fixing screws is prevented by the use of a back plate.

At low level, rim locks can be forced by a sharp blow from the foot on the door, opposite the lock. This will usually push the staple to one side (A, fig. 13.17). A man can bring more force to bear on a door by charging it with his shoulder. It is reported by the California Crime Technological Research Foundation that a man who is 6 ft tall and weighs 180 lb can produce shoulder impact against a door of 1800 lbf (8000 N) compared with 777 lbf (3450 N) with his feet (Stroik, J. ed. (1981) *Building Security*. Philadelphia: American Society for Testing and Materials. Atlas, R. 'Crime prevention through building codes', 88—97). Staples are also forced with jemmys. Screws through the staple into the door frame give increased strength, but failure occurs by splitting of the wood. A strengthened staple as shown in C (fig. 13.17) is necessary for strong security.

Other ways of attacking cylinder rim night latches are shown in fig. 13.17. D illustrates the familiar method of breaking a glass panel. Wood panels can also be broken or cut away to enable the latch to be reached on the inside. With a pair of footprints (combination pliers) the cylinder can be turned, as in E; cutting away the outer ring first allows the footprints to obtain a good grip. The method shown in F works because the body of the cylinder is brass. Thus the body can be pulled away from the steel fixing bolts by stripping the treads in the softer metal. Night latches must not be used on pairs of doors that do not have rebated meeting stiles, or elsewhere where a straight joint permits the insertion of a credit card or similar piece of plastics or thin metal (G). Good quality locks have pin

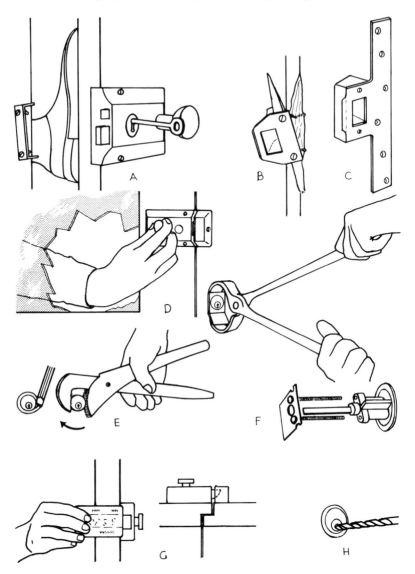

Figure 13.17 Attacks on rim locks and cylinder rim night latches. (A) and (B) Failure of staples; (C) strengthened staple; (D) deadlocking is a defence against this; (E) rotating the cylinder; (F) stripping threads in brass lock body, the pull plate aids the attacker; (G) credit card is effective in straight joint, stopped by rebate; (H) drilling out pin tumblers.

tumblers of hardened steel to resist drilling, as shown in H. (Similarly, mortice locks have a hardened steel plate over the bolt stump to prevent this being drilled out.)

Mortice locks

Cutting a mortice in a wood door for the installation of a mortice lock weakens the door. A burglar may take advantage of this to attack the door with a sledge hammer at the point where the lock is positioned, so breaking the lock out. A more subtle approach (fig. 13.18) employed when the mortice for the bolt of the lock is not far from the surface, as with an outward opening door or a pair of doors, is to cut away the material in front of the bolt. If the burglar starts by drilling holes, the work is speeded up and the break-in is accomplished with less noise. An L-section strengthening plate, as shown in C, should foil this sort of attack.

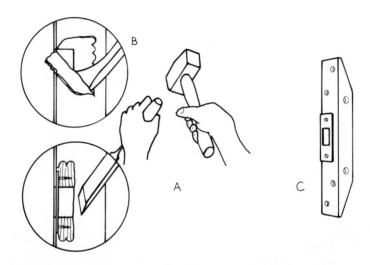

Figure 13.18 Cutting away jambs or meeting stiles to release bolt of mortice lock. (A) Wood jamb; (B) aluminium jamb; (C) strengthening plate.

Narrow-stile glazed metal doors pose a special security problem because of the narrowness of their stiles and because of their flexibility. The normally operated bolt of a lock narrow enough for the stile to accommodate may not have sufficient throw to stop it being prised out of its striking plate (fig. 13.19). For this reason locks with swing bolts have been developed. The actuating mechanism must resist being knocked down, or up, according to the way the bolt operates. The lock should also incorporate the other features shown to resist cylinder pulling and hacksawing.

Door opening limiters
Door chains allow a door to be opened slightly, for an occupant to speak to callers, and still restrict entry. However, a determined criminal can wedge the door open and cut the chain with bolt croppers. When a furtive entry is made, once the lock has been overcome it is possible to release a door chain of the type shown in A, fig. 13.20, with a bent coat hanger and a rubber band. The door limiter shown in B cannot be released in this way, it is designed to

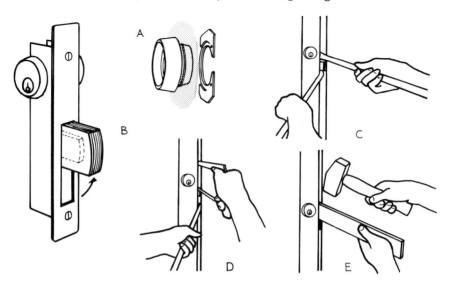

Figure 13.19 Narrow stile metal door locks. (A) Loose, bevelled ring and cylinder guard to prevent cylinder pulling (copyright Adams Rite); (B) swing bolt for maximum throw to resist prying as at C; (D) bolt of hardened steel or with alumina-ceramic insert resists hacksawing; (E) bolt must resist knocking down.

withstand a steady load of 7.5 kN (0.75 tonf) and is resistant to bolt cropper attack.

The position of letter plates in relation to all door fasteners must be carefully considered. If the reach of a hand (C, fig. 13.20) is not sufficient it can be extended by tools or special implements to turn keys and release other fastenings.

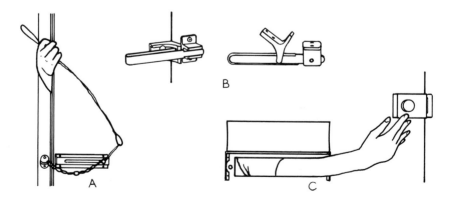

Figure 13.20 (A) Releasing door chain with wire and rubber band; (B) door limiter (Chubb); (C) letter plates should be at least 400 mm (16 in.) from any lock or fastener.

13.8   The selection of locks

The lock is only part of the security of a doorway, the construction and fixing of the door and its frame and the construction of adjacent elements of the building should have first consideration. However as all the items that ensure the security of a doorway are inter-related, the selection of the lock can seldom be entirely disregarded until the construction is settled. All external doors lockable from inside should be provided with bolts or supplementary dead locks. In selecting a lock for a final exit door, the form of attack likely in respect of the location of the door must be considered. BS 3621 (1980) *Specification for Thief Resistant Locks* provides a guide to the essential security features to look for. Briefly, the requirements of BS 3621 are that a lock:

(a) Has a minimum of 1000 differs.
(b) Must be resistant to operation by keys similar to the correct key.
(c) Must be resistant to drilling from the outside of the door.
(d) Must withstand a steady force of 13 000 N (3035 lbf) applied in a direction perpendicular to each of the faces of the lock.
(e) Must withstand the bolt being shot and withdrawn, by the key, 60 000 times.
(f) Must withstand steady pressure against the end of the bolt of 1200 N (270 lbf)
(g) Must have a bolt that will withstand hacksawing for five minutes.
(h) Will pass examination for general vulnerability by a panel of experts appointed by the Master Locksmiths' Association.

For basic security the minimum requirements are (a), (b), (d) and (f) of this list. Offical recommendations, such as the British Standard in preparation at the time of writing, *Security of Buildings, Part 1, Dwellings*, for the exit doors of dwellings, especially final exit doors, are for full compliance with BS 3621. For strong security this is a minimum requirement, for extra strong and maximum security a lock with increased resistance to picking should be selected and multi-point locking employed. The anti-drill plates of some locks are separate items, care must be taken to ensure that they are fitted. For all-round security, the strength of the door on the hanging side must be at least equal to that on the side where the lock is installed.

13.9   Key systems

There are two principal types of key systems:
(a) Master key systems
(b) Central locking systems
In a master key system, one key (the master) can be used to operate a number of different locks that are otherwise operated by individual (servant) keys. In a central locking system a number of different keys can be used to operate a central lock. The two systems are shown combined in fig. 13.21. Generally the systems are separate, master key systems being used for hotels, offices, factories, colleges,

etc., and central locking systems being used for blocks of flats. One use for a combined system is found in a private hotel; guests can let themselves into the building and their own rooms, while the proprietor holds the master key.

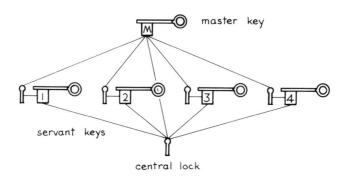

Figure 13.21 Combined master key and central locking system.

Both master key and central locking systems are capable of elaboration and further combination. Sub master keys can be introduced, say, to open all doors on one floor. If another rank of master key is required, as shown in fig. 13.22, the key that passes all locks is promoted to grand master. These hierarchical systems are useful where the structure of the organisation using a building or buildings limits staff responsibility to particular areas. The names of the keys are not standardised, what are here called master keys in a hierarchy under a grand master key are sometimes known as sub grand master keys. Further enlargement of a system, perhaps to cover more than one building, may require a great grand master key. In a central locking system more than one central lock may be provided: in a block of flats all residents may need to be able to open the door of the laundry and perhaps a side door in addition to the main entrance door. A central lock may be introduced into

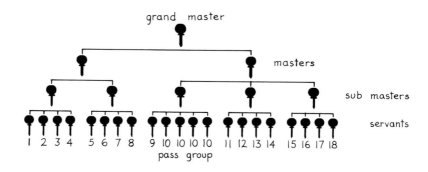

Figure 13.22 Master keyed suite with pass group.

part of a master key system with two or more keys operating it. There are numerous possibilities but it is best to keep any system as simple as possible, when a lock can be opened by more than one key a loss in security is unavoidable.

Some of the various methods of adapting lock mechanisms to master keying are shown in fig. 13.23. Lever and pin tumbler mechanisms can be made so that they operate when one or more levers or pin tumblers are lifted higher by the master key. In a lever lock a wider gate is necessary, this reduces security because it is less essential that the levers are lifted a precise amount. It is possible to avoid using the wide gate by having levers long enough to extend over two keyholes, one for the servant key and one for the master key, however this is an expensive solution reserved for specially made locks. The master keying of pin tumbler locks by dividing one or more pins reduces the number of ordinary differs. This leaves a sufficient balance for most purposes but, if necessary, the number of pin tumblers may be increased. Keys for Kaba locks can be drilled with different codes on the same key which means that the security of the various locks the key operates is not impaired by master keying. In some master key systems, locks have removable cores which can be changed if a key is lost or staff have to be replaced. To remove the cores a control key is used as shown for the 'Best' system in fig. 13.23, thus skilled people do not have to be brought in to change locks. This has the additional advantage that by the use of 'contractor's cores' during construction, the building owner can install his own system on taking over.

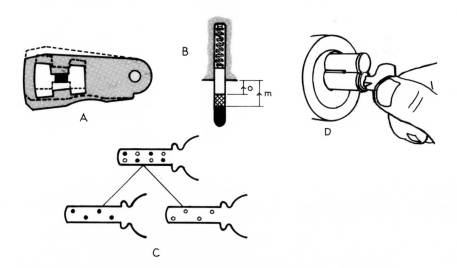

Figure 13.23 Master keying. (A) Dotted outline shows how master key lifts lever higher; (B) division in pin of cylinder lock, o — lift of ordinary key, m — lift of master key; (C) simplified representation of how codes for two locks can be drilled in a Kaba key; (D) removing 'Best' core with control key.

13.10   Electric strikes and electric locks

Electric strikes are used with ordinary locks in place of normal striking plates. When the latchbolt or dead bolt of the lock is shot, the electric strike holds the door closed without affecting the operation of the lock by key or inside handle (fig. 13.24). When electrically actuated via a control unit operated by a push button or other device, the strike releases the bolt, hence its alternative name of electric release. Electric motorised locks can operate without impairment of their manual operation, though it is a matter of choice, having regard to the function of the lock, whether or not key or handle operation is incorporated.

Table 13.2 summarises the principal ways in which electric strikes and locks perform. Both may be solenoid operated; in strikes the keeper is restrained mechanically in the closed position by a method such as that shown in fig. 13.24. In locks the latchbolt is held out of the lock case by a spring in the usual way, and withdrawn by the

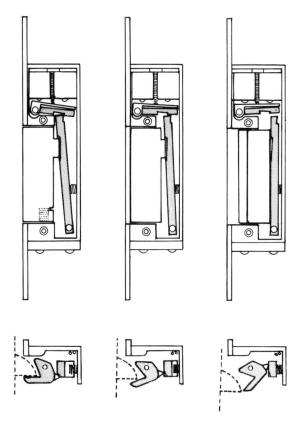

Figure 13.24 Electric strike or release, with side cover removed (Eff Eff). When current passes through the solenoid the top lever is attracted upward releasing the side lever and thus allowing the release of the door latch.

Table 13.2 Electric strikes and locks

Type	Operation	Control	Equipment
Intermittent	Unlocked for short periods to allow entrance or exit. Open when carrying current, locked when no current.	Impulse switch, e.g. push button, card reader.	Solenoid operated strikes. Electric motorised locks with latch bolts.
Continuous	Unlocked for long periods, such as during working hours, or opening hours of automatic laundrette. No current required to maintain locked or unlocked.	On/off switch. Automatic timer.	Reverse action solenoid operated strikes. Electric motorised locks with dead bolts.
	Locked in an emergency to bar entrance.	Personal attack button.	
Fail-safe	Locked when carrying current. Open when no current.	Impulse switch. On/off switch. Automatic timer.	Reverse action solenoid operated strikes. Electro-magnetically operated strikes. Solenoid operated locks. Electro-magnetic locks.

solenoid; alternatively the latchbolt may be withdrawn by an electric motor. Dead bolts are moved out and back by solenoids or motors.

Solenoid operated strikes and locks do not always release when under pressure, as from a door closer or someone pushing against the door, which could be serious in an emergency. Strikes operated by electro-magnets are safer in this respect, but much more expensive. A solenoid operated lock designed to release under pressure is shown in fig. 13.25.

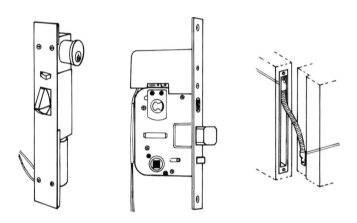

Figure 13.25 Electric motorised locks. (A) Solenoid operated, narrow stile lock with shaped, pivoted bolt for fail safe release from inside; without power the bolt of this lock is normally retracted (Kam-lok); (B) electric motor operated lock with latch bolt (Abloy); (C) lead cover for protection of cable from frame to door.

Electric motorised locks can give greater security than electric strikes, they can have a longer throw than that which an electric strike can accommodate and the throw can be into a boxed striking plate in contrast to the unsupported keeper of the electric strike. To protect electric strikes from vandalism a steel shield, large enough to cover the strike, should be fixed to the door by through bolts.

Different AC or DC voltages may be used to power the strikes and locks. AC strikes that have an inherent vibration and buzzing when the solenoid is actuated may be useful for releasing doors under pressure and alerting the user to their operation. For emergency purposes battery power can be used, if necessary circuitry being provided to operate AC strikes with DC current. Some locks have the control governing the time the door remains unlocked in the lock case, other locks and strikes have a separate control unit. A sensor microswitch, fitted into the strike, striking plate or lock face-plate, is used when it is required to monitor whether the door is closed and the bolt shot. Unfortunately this can be made to play false by the simple expedient of holding in the sensor. Another sensor activated by the solenoid of an electric strike may be used to indicate whether the keeper is open or closed. One way the microswitches can be used to monitor the status of a door and strike is to use coloured lights as follows:

(1) Secure, green light: bolt in strike, keeper closed.
(2) Released, yellow light: bolt in strike, keeper open.
(3) Open, red light: bolt out of strike, keeper open.

A tool attack test is included in ANSI (American National Standards Institute) UL 1034 (1974, 1980) *Standard for Burglary Resistant Electric Locking Mechanisms.* The product is mounted as in service and the attack carried out for five minutes by an operator familiar with its construction. The product must resist being opened with hammers, chisels, adjustable wrenches, pry bars, pincers, screwdrivers, picking tools and wires. The hammer is limited to a maximum head weight of 1.4 kg (3 lb) and no tool may exceed 457 mm (18 in.) in length. At the conclusion of the attack the locking mechanism must operate as intended.

13.11 Lock operation without keys

Door control without the use of conventional keys is accomplished principally by push button or card actuation of mechanical locks and by push button, switch or card reader operation of electric strikes and electric locks (fig. 13.26). Gaining entry by pressing numbered buttons in a coded sequence can be a lot less troublesome than the use of a key when frequent entry through a locked door

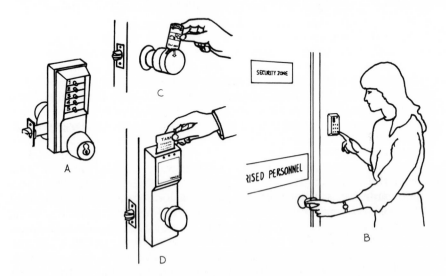

Figure 13.26 (A) Mechanical digital lock (Unican); (B) electronic digital release (Newman); (C) card activated mechanical-magnetic lock (Corkey); (D) card activated electronic lock (Yaletronics).

is required, as perhaps between public and private parts of a building. A digital coded lock or control unit is also useful where the use of an ordinary lock would require the issue of a large number of keys.

Card actuated locks and control units can utilise many more detainers (mechanical and electronic) to keep the bolt in position than conventional locks can use levers or pin tumblers. Not all the detainer positions have to be used in any lock or unit, so involved master key systems are comparatively easy to arrange. Additionally, combinations (mathematically they are permutations) can be changed with ease, in some mechanical systems without dismantling the lock, in other systems by electronic control. Frequent changes of combination are essential where access has to be made available to a changing population, as in a hotel or students' hostel, and where there is a large staff turnover: in the U.S.A. about 80% of hotel burglaries are made by use of illegally acquired keys.

In radio systems of lock operation without keys, the user carries on his person a miniature transmitter which sends out a signal with a range of two or three metres. The signal is picked up by a sensor controlling the lock. Another system checks the user's hand geometry. It can be used alone or, for greater security, in combination with a digital code or card reader. The hand is placed against a sensing plate that reads the geometry of the hand and verifies that it matches either information on an access card inserted at the same time in a card reader, or information previously entered into a computer.

Access control systems using solid state electronics may be linked to a central controller to give what is known as an 'on-line' system. The controller makes the decision whether or not to allow access according to the way it has been programmed — not to allow entry outside an employee's working hours, for example.

Whatever control unit is used, the lock actuated should be electrically motorised if strong physical security is required, otherwise a solenoid operated lock or an electric strike may be used.

In many locations it is advisable to fit a door closer to ensure that the door closes after each operation. With autonomous or 'off-line' systems it may be thought necessary to install a buzzer or other device to give a warning that the door has not been closed after an entry has been made; on-line systems should indicate this on the controller.

Digital code actuation

The push buttons on the mechanical digital lock shown in A, fig. 13.26, are shielded to make it difficult for a would-be intruder to read the code as it is punched in. Nonetheless he may read it or learn it from a user. The code can be changed easily so this should be done whenever it is felt that security might be compromised. A cylinder lock allows the push button operation to be by-passed by the use of a key in the cylinder. This is a convenience for management use when codes are changed or when several locks are in use each with a different code.

An electrical digital control will usually be mounted at the side of the door as shown in B. A nine digit push button unit such as this is normally programmed to operate in response to a four digit code. This gives over 3000 permutations to select from. Someone might stumble on the chosen code by randomly, or systematically, pushing buttons; to preclude this control units may incorporate a short duration timer to disallow further entries for three or four seconds after a wrong code has been entered. Codes may be changed by interchanging colour coded jump wires within the unit, or in more sophisticated units by turning a key-operated switch and entering the new code by pressing the appropriate buttons: the effect of turning the key is to store the new code in the solid state memory. The time for which the electric strike is actuated can be set for up to about thirty seconds, when the equipment is installed. Any suitable lock with night latch operation can be used, knob or lever handle operation of the latch will be required on the inside of the door. Manual operation of the lock, as normally used on leaving, will be essential in the event of failure of the power supply or if burglars or vandals render the control inoperative, because in such circumstances the door will remain locked. The use of a key from outside at any time will operate the lock in the usual way, over-riding the electric control.

Card actuation
The card actuated, mechanical-magnetic lock shown in C, fig. 13.26, has a locking mechanism that contains a number of magnetic pins in wells in a plastics core. These pins are attracted to a steel plate through holes in a brass locking plate. While under attraction the pins are still partly in the plastics core, thus the mechanism is locked. To release the mechanism a plastics card with magnetic spots in the material within the card is inserted into a slot behind the steel plate that is attracting the pins, the magnetic spots repel the pins back into the plastics core enabling this to be moved round to engage the bolt operating mechanism. Stainless steel cards can also be used. Both types of card are coded using an electronic encoding gun; the cards are erasable with degassing equipment and can be re-coded when combinations are changed.

In another form of card access the equipment compares the user's card with a master card held inside the lock or reader. If the codes on the two cards are compatible a number of small magnets are moved enabling the user to insert his card farther into the slot. This operates a microswitch which in turn activates the circuitry that opens the door. All, or selected, user cards can be made invalid by changing the master card. This makes the system of special value to clubs, members have a convenient means of access to the premises coupled with a membership card and those who have not paid their subscriptions can be excluded.

The lock shown in D, fig. 13.26, is a unit in an advanced electronic system designed to produce a high level of security for hotel guest rooms. No wiring is required. Every guest is allocated a unique code, and a card is punched accordingly by a central computer allied to a card punch. Within the lock is a microprocessor synchronised with the central computer, this will only open the lock in response to the insertion of a correctly-coded card. Endless guest codes can be produced together with master codes at five levels, and the issue of a new code automatically cancels all prior codes at that level. When a guest leaves his card is discarded — he can keep it if he wants to. If a repairman needs to enter a room a card can be produced to permit one entry only.

On commercial premises, sophisticated on-line systems with computer controllers can be used to verify a card-holder's authority for access regarding time and place, locate a person by giving a 'last door used' report, display the last hundred or so transactions, and with card reading on leaving as well as on entering, clock the attendance of employees working normal and variable hours. The controller can also be used for purposes other than the coming and going of people, for turning off lights, for example. From a safety aspect, it can inform rescue services who is in a particular area in the event of an emergency.

For tight access control in commercial premises, research laboratories and similar establishments, card readers can be combined with push button control. Users are required to enter a personal identification number as well as use their card. Thus a stolen card is not sufficient to gain access. If a card holder is forced to grant entry under duress he can type in a special code to actuate a silent alarm.

An example of how an access control system using card actuated locking devices to control the passage of vehicles and people is designed is illustrated diagrammatically in fig. 13.27. There are different levels of access for different parts of the building (controlled, limited and exclusive, see section 3.4) but card holders use a single card for all access authorised to them. In the car park, low grade security with no control of people on foot is acceptable because the

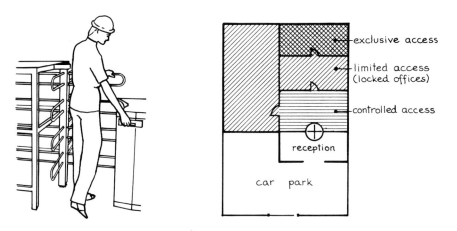

Figure 13.27 Access control, Use of card reader (Plantime) and diagrammatic representation of varying levels of safety.

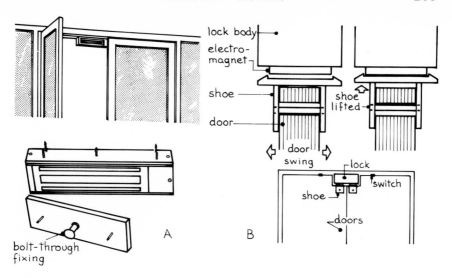

Figure 13.28 Electro-magnetic locks. (A) Lock with face-to-face contact (Warshaw); (B) lock with deadlocking armature plate (BSG Security).

object is merely to prevent unauthorised parking. Controlled access to the building is through a turnstile or revolving doors that allow only one person through at a time: this is to prevent two people entering with one card. The card holder then operates the door to his own office or workplace, which has limited access; only if his work or position requires him to visit other parts of the building will he have a master card giving access to other offices. Finally, entry to the exclusive access zone requires personal identification by means of a digital code in addition to use of the card.

13.12   Electro-magnetic locks

Also operated by electric control devices are electro-magnetic locks. Quite different from other locks in their mode of operation, most locks of this type have no moving parts, they consist simply of an electro-magnet and a steel plate. The electro-magnet is fixed on the head of the door frame, and the steel plate on the door (fig. 13.28). When the electro-magnet is energised magnetic attraction holds the steel plate firmly to the magnet thus 'locking' the door. The power of attraction is sufficient to resist a force in excess of 4500 N (1000 lbf). It is a fail-safe lock because it operates only when energised. When wired to a fire alarm it is suitable for emergency exit doors. If desired the lock can be fixed to the jamb of the door frame instead of the head, it can also be used with sliding doors.

Though electro-magnetic locks have no moving parts to jam or wear, residual magnetism can cause sticking. Precautions are taken in the design of the locks to minimise this and to counteract back e.m.f.

harmful to the equipment. The strike plate, or armature plate, may be nickel plated and the magnetic core gold plated to give protection against rusting. Micro switches, wired to flashing lights or an audible alarm, will indicate when the door is open. Alternatively the lock can monitor itself by means of a voltage signal that indicates when the door is secure.

Whether the power of attraction is sufficient to resist a shoulder charge is questionable if it is accepted that a man can exert a force of 8000 N (1800 lbf) in this way (section 13.7). Thieves also force magnetic locks by 'peeling': they use wedges to free the door gradually from the grip of the lock. If a sticking plaster can be previously stuck on the strike plate it assists by reducing the holding power of the magnet. One kind of electro-magnetic lock has an armature plate in the form of a 'shoe' as shown in B in fig. 13.28. The lips at the end of the shoe add mechanical strength to the magnetic attraction: an elaboration that introduces a moving part but counteracts peeling. Owing to the relationship of the electro-magnet and the armature plate this kind of lock is particularly suitable for swing doors with double action.

13.13 Phone entry systems

Used principally for the avoidance of vandalism in blocks of flats, phone entry systems are also used to control access to security areas in other buildings. The visitor indicates that he desires entrance either by pressing a button against a named display or by entering a code on a digital call panel. The person receiving the call speaks to the caller through a telephone handset or microphone-speaker unit and permits entrance, if this is acceptable, by pressing a button that actuates an electric strike or electric lock, so releasing the entrance door. High quality audio equipment is essential for voice recognition. Phone-plus-TV systems (fig. 13.29) give more positive identification at increased cost. Systems generally may have a single user's unit, as when an entrance is controlled from a remote office, or hundreds

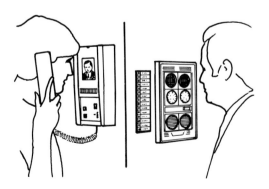

Figure 13.29 Entry phone-plus-TV.

Figure 13.30 Vandal attacks that phone entry panels (and lift control panels) must resist.

of user's units, as in a large block of flats — normally an upper limit of about fifty is desirable. Where there is a porter or commissionaire a porter's unit with a switch-over facility may be provided so that the door can be controlled from individual user's units or the porter's unit — useful when tenants are away or offices unmanned. Extra security is given by the capacity of this system for communication between individual users and the porter when necessary.

With a named display of occupants or flat numbers an intruder may try to gain entrance by pressing all the buttons in succession and hoping that someone will release the door without checking on the caller. Digital push buttons that require the flat number to be entered make this trickery more difficult (and need less wiring), greater security is given if coded addresses are used but a visitor will need to know what code to use, or call the porter if that is possible.

Vandal resistant controls, as shown in fig. 13.30, are essential for most installations. They should have all-metal fronts with non-protruding buttons to resist blows and ensure that the shock is taken by the switch body, not the contacts. Plastic buttons should not be used because of their vulnerability to damage by flame from a lighted match or cigarette lighter. Additionally the gap between prezzel and bezel of push buttons must be as close as possible to prevent the insertion of a knife point, or jamming by a match.

Chapter 14

Windows

14.1 The problems

Were it not that windows are glazed with a fragile and dangerous material and are made openable to act as ventilators, the only safety and security problems with windows would be their safe cleaning and maintenance. Glazing with safety glass or safety plastics can remove the danger of cuts and lacerations to people impacting with the window, but not necessarily the danger of them falling through broken glazing: the material and glazing system must be properly chosen for this purpose (see section 6.5). Security is not greatly aided by alternatives to annealed glass unless a very tough material is used, such as polycarbonate, which may not be aesthetically acceptable. Ventilation by separate ventilators so that all lights of a window are fixed would keep in the people who fall from open windows and keep out the burglars who force an openable light to gain entry. But in domestic and similar buildings, ventilation solely by separate ventilators is usually impracticable. Anyway, people like to open windows to get a feeling of contact with outdoors in fine weather.

When a window must be located in a position where there is a risk of impact, at the foot of a stairway, for example, as in fig. 14.1, safety glass or safety plastics must be used (see also section 11.9 for window hazards on stairways). When a window has to be in a vulnerable security location then it must have fixed lights only; the opportunist and minor criminal will not usually smash a large pane and climb in over broken glass, his object in breaking a pane is more often to reach the window fastener. For strong security either the glazing must be capable of stopping entry or some additional protection, such as bars, must be used.

SAFETY

14.2 Struck-against accidents

Just as people walk into glazed doors not realising they are there, so people walk into windows and glazed wall openings not realising

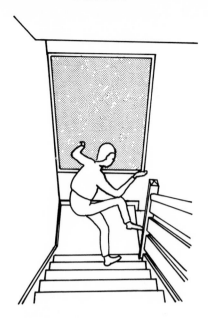

Figure 14.1 An obvious but not uncommon danger. Annealed glass in such a position is not commensurate with safety.

that they are glazed. BS 6262 (1982) *Code of Practice for Glazing for Buildings* allows annealed glass to be used in windows from floor level up to any height in areas subject to the movement of few people. This is not considered acceptable by the Child Accident Prevention Trust, who advise safety glazing wherever glazing extends into an area 300–800 mm (1 ft–2 ft 7½ in.) from the floor. In areas subject to the movement of many people, annealed glass can be used at all levels and still comply with BS 6262 if a protective rail of 75 mm (3 in.) is fitted in front of the glass. Such a rail will not give acceptable safety, people are likely to attempt to open what they think is a glass door by pushing on the glass, particularly if they have entered through a glass door. They can strike the glass heavily if walking rapidly towards it at the time. If a rail is considered necessary then safety glass or safety plastics should be used, the rail will then denote the presence of the glazing and stop people walking into it.

Children can come into contact with annealed glass in windows in many ways that a protective rail would not stop (see fig. 14.2 for example). Along corridors and in assembly rooms, where BS 6262 permits annealed glass in the danger area of within 800 mm (2 ft 7½ in.) of the floor, providing it is protected by a rail or by being recessed; the designer must consider whether in such locations a rail or recess will give adequate protection not just to people using the corridor or assembly areas but to people below should the glass be broken (fig. 14.3). Upper storey rooms of houses are not regarded as routes of travel so protective barriers are not usually considered

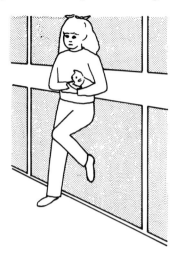

Figure 14.2 A girl, bored while visiting family friends with her parents, rested her foot on low level glazing, her foot went through the glass and she was cut on the leg.

necessary, nonetheless safety glazing should be provided, for the protection of children at play, if for no other reason.

Struck-against accidents occur in corridors when people walk into the edge of an open window (fig. 14.4). Pivot hung and inward opening windows must not be used on traffic routes.

14.3 Openable windows

Openable windows lead to accidents when people have difficulty

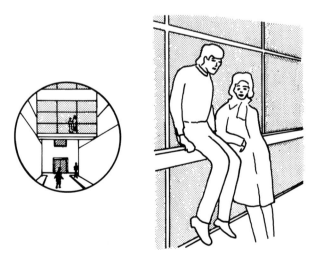

Figure 14.3 Even if not misused, a rail does not give sufficient protection for low level annealed glass, and broken glass may find victims below.

Figure 14.4 Repetition of windows on the elevation put this openable window in a dangerous position on the stairway; a fixed light should have been specified.

opening them, when they open unexpectedly, when people lean out too far and when people have to stand off the floor or over-reach to open them. A hard push on an outward opening window that is stuck frees it more successfully than anticipated, and the victim overbalances and falls out. A child playing on a window ledge, hiding behind the curtain perhaps (fig. 14.5), or just looking out (fig. 14.6),

Figure 14.5 Children have fallen to their death playing hide-and-seek behind curtains.

Figure 14.6 Only a bulky nappy saved a fourteen month old boy from falling twenty-two storeys from a window like this that was open for ventilation.

leans on the window that has not been fastened properly. A man parks his car under the high-rise flat where he lives, leans out to see if it is all right and falls to his death. An elderly woman stands on a stool to open a window out of reach, or reaches over a cooker to get at the window catch, falls off the stool, or tips over a saucepan on the cooker (fig. 14.7). These are the kinds of accident that design for safety can prevent.

Openable windows must be considered for safety principally from the point of view of a child. Falls from windows are responsible for about 1% of all fatal accidents to children. Many children fall and survive, some from as high as six storeys, but are left with crippling injuries. Boys and toddlers are most at risk. To make openable windows as inherently safe as possible, the following precautions must be taken (see also section 2.7):

- *Permanently* restrict the distance the window can be opened to 100 mm (4 in.), or
- Make the height to the bottom of the openable window well above that which a young child can reach by standing on a chair or table, say 1350 mm (4 ft 5 in.).
- Ensure that an inner sill or window ledge does not provide a platform for a child to stand on.

For a two-storey house there may be objections to the permanent restriction of window opening and to high-level, openable windows because these two measures will not allow escape in case of fire. When there is a ground floor fire, smoke and flames flow up the stairway. The burning materials of furnishings — wood, wool, plastics and paper — give off carbon monoxide (CO) and carbon dioxide (CO_2). In addition the burning of polyvinyl chloride (PVC) in wall coverings and upholstery produces hydrogen chloride (HCl); polyurethane foam, as used in modern furniture, produces hydrogen cyanide (HCN). Small proportions of these gases in air can be fatal but carbon monoxide is the greatest killer: a few breaths will render a person unconscious; death occurs in minutes. Once furniture in a downstairs room is blazing fiercely with the room door open or burnt through, people upstairs have two to five minutes in which they can possibly escape.

Figure 14.7 Special care is necessary in positioning kitchen windows; if they cannot be reached in safety window controls are necessary.

In a house in Birmingham where the opening lights of the upstairs windows were 1.5 m (5 ft) from the floor, with fixed panes underneath, a mother and her three young sons died in a fire. They were trapped upstairs and asphyxiated by smoke and fumes. If they could have opened the lower part of the window they might have escaped on to a flat roof below. Neighbours saw the mother banging on the window. She might, perhaps, have used a chair or some other object to break the window glass. However, before the physiological effects of carbon monoxide and other products of combustion are present in sufficient quantities to cause death, their presence can cause a person to act irrationally and with impaired judgement, so that the person fails to take sensible steps to save himself. In any case, breaking a pane of glass deliberately is sometimes surprisingly difficult and, when it is broken, jagged pieces remain to cause serious injury.

Though impromptu rescue with ladders owned by neighbours is not uncommon in house fires and though we often hear of people saving themselves by jumping from first floor windows, whether these actions are warranted in the circumstances is usually unclear. Stress coupled with impairment of judgement brought about by fire gases can occasion desperate action; doubtless many who have jumped from windows could have waited safely for rescue.

It is no way out of the falls-escape dilemma to make the windows above the flat roofs of porches, garages and outbuildings open easily with a clear space for exit. The danger to a child falling from the window is lessened but an easy way out for occupants is an easy way in for burglars. The designer must consider where the greatest risk lies: usually falls are more likely than life-endangering fires. If, however, a window is made so that it can operate as an escape route, the risk of burglary may be reduced by fitting it with an intruder alarm. A self-contained 'mini-alarm' will give warning whenever the window is opened, by child or burglar.

14.4   Cleaning and maintenance

In England designers make provision for cleaning domestic windows from inside at third floor level and above, but usually assume that householders will be able to get window cleaners to clean the windows at ground, first and second floor levels. Actually it is difficult to get the one- or two-man firms who undertake domestic window cleaning to work above the first floor — they do not carry ladders that are long enough — and occupants are driven to take risks in cleaning the windows themselves (fig. 14.8).

Figure 14.8 A young woman was killed attempting to clean a window in this way when the hinge broke.

In Scotland the Building Standards (Scotland) Regulations require that windows in houses and flats must be capable of being cleaned from inside the building when any part of the window is more than 4 m (13 ft) above the ground. An exception is made if cleaning is possible from balconies, flat roofs or permanent platforms, properly guarded. With buildings other than houses or flats, cleaning may be from a suspension system or travelling ladder system, or up to 9 m (29 ft 6 in.) high from a portable ladder provided there is adequate unobstructed space for its use.

Cleaning by occupants

The apparently simple solution to the problem of cleaning and maintenance of the outside of a window from the inside by occupants is to use side hung, inward opening lights. However, these necessitate special precautions to prevent the entry of driving rain, there is also the drawback of the way rain falling on a window open for ventilation runs into the room. As with any window that has to be made so that it can be opened more than 100 mm (4 in.), an inward opening light is not inherently safe as reliance has to be placed on the correct operation of a safety stay. The problems associated with safety stays have been appraised in section 2.7: their principal shortcomings are that their effective operation is likely to be short-lived especially if they are not subject to regular maintenance, and they are likely to be over-ridden when occupiers feel the need for abundant ventilation.

Side hung, outward opening lights may be fitted with hinges of the 'easy cleaning' type (fig. 14.9). These have an offset pivot so that when the window is open there is a gap through which the cleaner's arm may be inserted. The minimum distance between the window frame and the open light should be 95 mm (3¾ in.) according to BS CP 153 (1969) *Windows and Rooflights Part 1: Cleaning and Safety* and this is only satisfactory when the window frame is fixed not more than 100 mm (4 in.) from the external wall face. Even if the cleaner can get an arm through the gap, the arm will be in danger of abrasion from the reveal if that has a rough finish such as pebble dash or roughcast rendering. Wind pressure that tends to close the window will tighten it on the cleaner's arm, therefore there must be some means of fastening the window securely when open at 90°.

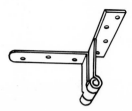

Figure 14.9 An easy-cleaning casement hinge; this type should not be used where burglars can remove the fixing screws.

The cleaner's reach through the gap will not normally exceed 560 mm (1 ft 10 in.) when a cloth is used and many people will not achieve this if they have to raise their arm above shoulder level. A 'squeegee' on a long handle makes the job easier and safer but occupants cannot be relied upon to obtain one and use it, they are likely to attempt to reach parts inaccessible from inside by leaning out from the open window. Many people do not have the physique to reach more than a small part of the window even if they are prepared to put their safety in jeopardy (fig. 14.10).

Figure 14.10 Easy-cleaning hinges are seldom the solution to the problem of safe, external window cleaning.

The unsuitability of the traditional casement window for cleaning the outside from the inside (in spite of easy cleaning hinges) and of the double hung sliding sash window (fig. 14.11) has led to the development of various other types of window. One of them, the reversible horizontally pivoted window was appraised for safety in section 2.7. Unfortunately, in spite of the problems that have arisen with this type some manufacturers are still producing windows that can be opened more than 100 mm (4 in.) for ventilation. Also, in addition to the shortcomings already revealed, some windows will not reverse completely; this is most undesirable especially when it means that the window cannot be secured in the reversed position.

There are windows with inward opening bottom hung lights that are turned into inward opening, side hung lights for cleaning, and outward opening, top hung lights that are turned into reversible

Figure 14.11 The traditional sliding sash window has a wide sill which reduces the risk of this basically unsafe practice. Thinner aluminium windows, set nearly flush with the external wall face have very narrow sills making this method of window cleaning very hazardous.

horizontally pivoted lights. Also there are horizontally sliding windows and vertically pivoted windows and projecting windows. All must be subject to a rigorous safety appraisal: safety devices must be viewed with suspicion and the ability of small children to release catches and fasteners should not be underestimated — when safety chains secured by a hook similar to that used on a dog's lead were fitted as a remedial safety device to windows in high-rise flats it was reported that the children, whom the chains were intended to protect, found it easier to release the hook than open the window. The easiest window to operate and clean safely from inside is the tilt and turn window (fig. 14.12).

Figure 14.12 A tilt and turn window (Combidur Systems). When open for ventilation it is restrained by a stay at the top. Children are safe as long as they cannot reach the release handle (which may be lockable when the window is closed).

To clean glazing adjacent to an openable light the maximum safe reach is given in BS CP 153 as 560 mm (1 ft 10 in.) sideways, 510 mm (1 ft 8 in.) upwards and 610 mm (2 ft) downwards. But the Code of Practice also states that the only fixed light that can be safely cleaned by an occupier through an opening light in a sheer wall face is a sub light not more than 600 mm (2 ft) deep. Another requirement is that, for second to sixth storey windows, where a person has to reach through an opening large enough to fall through to clean a sub light or part of an opening light, these should be within easy reach when the person is standing at least 1.02 m (3 ft 4 in.) below the opening, or safety rail if one is fitted. Above the sixth storey, because of the greater risk of vertigo with increasing height, the cleaner should be able to reach when standing 1.12 m (3 ft 8 in.) below the opening or safety rail, this is just above the centre of gravity of the ninety-ninth percentile male (see table 4.2).

Cleaning by professionals
Windows at high level may be safely cleaned outside from a suitable ledge or balcony. A railed balcony not less than 610 mm (24 in.) wide or more than 305 mm (12 in.) below the window sill exempts a window from ANSI A39.1 (1969) *Safety Requirements for Window Cleaning*, providing the railing is at least 1070 mm (42 in.) high. A minimum width of walkway of 635 mm (25 in.) is specified in BS CP 153, Part 1, *Windows and Rooflights — Cleaning and Safety*. Where there is no balustrade there must be facilities for the use of a lifeline and at least 230 mm (9 in.) of the walkway should have a non-slip surface. For a range of windows a track and roller unit will be necessary for the attachment of lifelines (fig. 14.13). The track can be of channel or rail section mounted on the soffit or wall face above the windows. Handholds at not more than 1.5 m (5 ft) apart, preferably at a height of about 1.2 m (4 ft), should be provided. It is advisable to limit the height of window to be cleaned to about 2.1 m (7 ft) above the walkway, the cleaner will be able to reach higher with a long-handled squeegee but he may be tempted to stand on something instead. Safe access to the walkway is an essential part of the design.

Where the length of the window to be cleaned is limited, a safety eyebolt or other form of anchor may be used for the attachment of the lifeline. Requirements for eyebolts, with recommendations for their fixing and testing, are given in BS 5845 (1980) *Permanent Anchors for Industrial Safety Belts and Harnesses*. Approval requirements for anchors and anchor fittings and methods of testing installations are given in A39.1. In the British Standard the recommended methods of fixing are to masonry and structural steel, A39.1 allows anchors to be fixed to windows.

For fixing eyebolts to masonry, through-wall anchors, cast-in sockets, expanding sockets and chemically anchored sockets are used. An inside fixing has the advantage that the eyebolt will be protected from the weather, also the lifeline can be attached before the cleaner is exposed to the risk of falling. A39.1 allows an outside

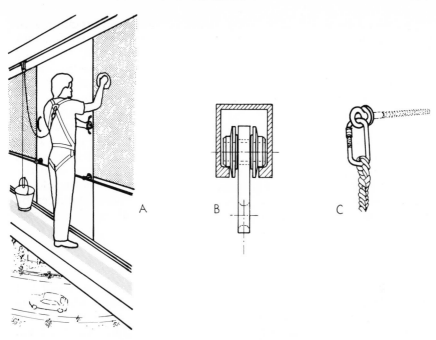

Figure 14.13 Lifelines. (A) Track system with handle and rail handholds; (B) channel and roller unit (Cradle Runways); (C) safety eyebolt.

fixing if it is possible for the cleaner to attach and release the lifeline by extending only an arm through the window. Generally the most suitable position for an eyebolt will be the inner reveal. Where the fixing is in a brickwork or concrete reveal there must be a thickness of at least 150 mm (6 in.) from the anchorage to the outer face of the reveal. Similarly, when the anchorage is in the inner wall face it must be at least 150 mm (6 in.) from the edge of the reveal. The recommended height above the sill for inside or outside fixing is approximately 1000 mm (40 in.). A39.1 specifies fastening anchors to the window frame or mullion 1070–1300 mm (42–51 in.) above the sill. The window frame must be in sound condition and be securely fastened in place, the manufacturer of the window unit being required to submit evidence that the complete installation has withstood a specified drop test without failure at the point of attachment of the anchors and without detachment of the window unit from the wall.

The American standard A39.1 permits the window cleaner to stand on the sill, subject to there being a certain minimum standing room in relation to the slope of the sill. However it might be felt that for safe working a proper working platform is necessary, even for a single window, and if the cleaner is compelled to stand on the sill holding on to the window frame his safety has not been provided for.

For long runs of vertical or inclined glazing, cleaning may conveniently be carried out from travelling ladders. Figure 14.14 shows a travelling ladder for a combination of vertical and inclined windows. Ladders in general may be permanently attached to rollers running on a track or may be coupled to the rollers when required. Tracks may be on roofs as in the figure or, if for vertical windows, fixed to the soffit or wall face. There must be a safe platform for mounting the ladder and provision must be made for the cleaner to use a lifeline, preferably one with an inertia brake.

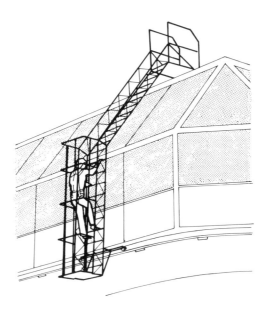

Figure 14.14 Travelling ladder with runway system at roof level and bearing track and base of ladder (Cradle Runways).

Suspended cradle systems are necessary for the cleaning of windows and general maintenance of tall buildings. Various systems are available but an examination of them, opening up the whole problem of the maintenance of tall buildings, is beyond the scope of this book. However mention might be made of the ultimate in window cleaning safety: the automatic system developed in the U.S.A. for the cleaning of glass fronted buildings. Up to the time of writing, only one automatic system has been installed in Britain, this is on the National Westminster tower in London. A washer head containing detergent sprays, agitator brushes and squeegees moves down the face of the building washing, wiping and vacuuming on its way.

14.5   Accidental and casual loading

Though cleaning and use of a window designed for safety do not require occupants to lean ladders against the window members or apply considerable force to them in other ways, the members need to be strong enough to withstand accidental and casual loading equivalent to, say, that which would occur if a man fell against a member. Straining against a member to pull a banner across a street is one example of casual loading; holding on to an open light when decorating from a ladder is another; children at play might apply forces in various unforeseen ways. In the Building Research Establishment Digest 262 a review of window performance shows that to withstand accidental loading it is authoritatively thought that a window should resist at any point, without breakage or permanent distortion, a perpendicular force of 500 N (112 lbf) applied without shock and held for one minute when the window is in a fully open position or against a restrictor device. When a force of 1000 N (225 lbf) is applied under the same conditions the window should not be expected to survive undamaged but it should not collapse and the restrictor device should not become detached. An impact energy of 13 kg (28.7 lb) falling 100 mm (4 in.) has been proposed for assessing the effect of a window slamming against a restrictor device.

SECURITY

14.6 Security weaknesses

Openable windows are normally less secure than doors and they are more often left open when they should be closed, as a result most break-ins are through windows. Even so windows are not a security risk unless they are accessible, therefore the number of ground floor windows should be restricted, and openable windows should be too small or too awkward for burglars to get through. Climbing routes leading to roofs and balconies where there are windows should not be formed (see fig. 8.14). Though the use of windows as an escape route, as described in section 14.3, should be considered for two-storey buildings, no such consideration is necessary for taller buildings as this use of windows is not acceptable for them.

Thieves can break in through windows by dislodging casement stays and fasteners. Without breaking the glass, this can sometimes be accomplished with a piece of wire, a knife or strip of thin metal or plastics pushed through the joint between the opening light and frame. With wood windows a hole made by drilling enables a wire or rod to be used to dislodge the stay or move the fastener (fig. 14.15). If the window is fairly newly glazed and the putties have not hardened fully they may be chipped out and a pane removed. Glazing gaskets may allow easy removal of the glass. Leaded lights can sometimes be pushed in (see fig. 6.2) and ill-fitting horizontal sliding windows lifted out of their grooves. Hinges are made that can be unscrewed from outside, some hinges have pins removable from outside. Burglars have removed extractor fans fitted into window glazing, put a hand through the hole and released the window fastener (fig. 14.15).

Glass louvre blades can be fixed in place with an adhesive to

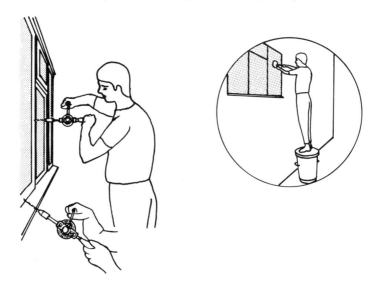

Figure 14.15 Drilling to release fasteners — a quiet way of forcing a window. (insert) Extractor fans and ventilators must not be removable from outside.

prevent them being removed from their housings, but this does not prevent them being cut close to the housing with a glass cutter and knocked out. The remedy is for each blade to be framed in an individual narrow-sectioned metal frame.

A measure of how window assemblies suitable for single or multiple family dwellings will resist attack by unskilled or opportunist burglars is given by the application of ASTM F 588—78 *Standard Test Methods for Resistance of Window Assemblies to Forced Entry Including Glazing.* First the assembly is stripped of screws, glazing beads and other mechanical fasteners that can be readily removed in five minutes, then attempts are made to open the window by hand by pushing, pulling and otherwise manipulating the opening light, the locking device is subjected to attempts to reach it with a knife or wire, and the resistance of the assembly to forcing is tested by static loading.

14.7 Window locks

Locks for windows have to be small because windows have narrow members, thus the locks are weak and easily forced. However they do prevent the burglar dislodging fasteners and stays and opening a window unobtrusively. They will not resist a determined forcible attack but they are another obstacle to overcome when the burglar breaks the glass to release the fasteners or attempts force with a jemmy.

A number of locks have been developed for different types and sizes of windows manufactured in different materials, some of them are shown in fig. 14.16. Mechanisms vary from those incorporating pin tumblers with keys in a number of differs to simple turn or screw mechanisms operated by keys common to all locks of one kind. The fitting of locks after manufacture can prove difficult especially with aluminium and plastics windows. Locks for windows should not be an afterthought, if designed into the window they can be embodied in the sections, giving them the security of mortice locks as compared with rim locks.

14.8 Grilles

Security grilles are preferably fixed on the inside of the windows. There they are protected from the weather and are not easily sawn through or pulled off with the aid of a vehicle; windows are more easily cleaned, and they can open outwards for ventilation. Grilles of expanded metal or mesh welded to an angle iron frame, as shown in fig. 14.17, screwed or bolted to the window frame, will improve basic security in industrial buildings. Ornamental grilles with wrought iron scrolls are weakened by the decorative work but deter the minor criminal. Strong bars as in C increase the appearance of solidity of monumental type buildings when fixed externally, thus they are fixed outside the window for that reason. Removable grilles, as shown in fig. 14.18, may be used where twenty-four hour protection is not required; for example the type shown in A may be used to improve the night-time security of a shop window. The hinged grille shown in C may be preferred for ease of maintenance but security will be less than that given by a permanently fixed grille of similar construction.

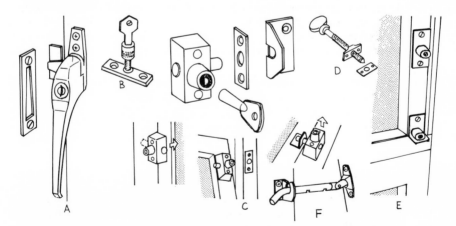

Figure 14.16 Window locks. (A) Casement fastener with cylinder lock for metal or wood windows; (B) casement stay lock, a security and safety device; (C) universal lock, and applications; (D) sash screw, used to secure double hung sliding sashes when closed or open for ventilation; (E) stop bolts, retractable into sliding sash, usually one fixed each side at different heights for security with open or closed window; (F) security and safety catch that can only be released when window is closed (P.N. Hardware).

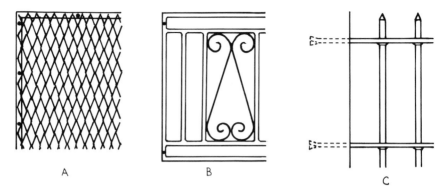

Figure 14.17 Fixed grilles. (A) Expanded metal; (B) ornamental; (C) barred.

Bars of grilles required to give strong to maximum security must be strong enough to resist spreading by jacks or other methods. Square bars should be 18 x 18 mm ($^{11}/_{16}$ in.) and round bars 20 mm (¾ in.) diameter at a maximum space of 125 mm (5 in.), centre to centre. Transverse bars of flat section need to be proportioned according to the width of the grille. Generally they should not be more than 300 mm (1 ft) apart.

In the U.S.A. bars may be required to meet the requirements of ANSI A627—68 (reapproved 1981) *Standard Specification for Homogeneous Tool-Resisting Steel Bars for Security Applications* and A629—77 (1981) *Standard Specification for Tool-Resisting Steel Flat Bars and Shapes for Security Applications.*

Though external bars may be built directly into the masonry, plastering and other finishings to the jambs are not likely to be well done when direct fixing is used inside. A more suitable method of fixing is shown in B where a hole is drilled in the jamb and a separate fixing bar is bedded in the hole. For a hinged grille the fixing bar is pre-welded to the upright bar; for a fixed grille the fixing bar is

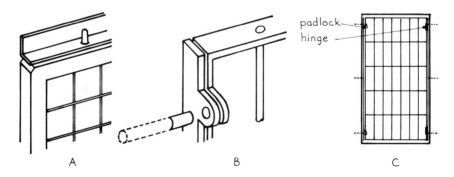

Figure 14.18 Removable grilles. (A) Grille held by pins at top, padlocks at bottom; (B) detail of staples for padlock, and fixing to joint, of hinged grille shown at C.

pushed through a hole in the edge bar, suitable measures being taken to prevent any sideways movement of a grille that is not a tight fit in the opening.

Collapsible grilles (A, fig. 14.19) may be used on the inside or outside of windows. Inside they are less vulnerable to attack but of course they do not provide the protection of fixed bars, their advantage is that they can be folded clear of the window when not needed. Outside they may be used as gates to give protection to entrances as well as windows. Grilles for domestic applications are available, when these are not in use they are folded out of sight behind curtains, and the bottom guide is removed to give an unobstructed sill.

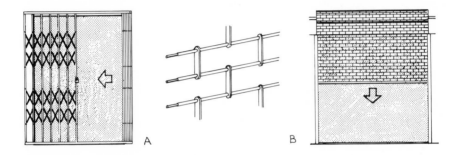

Figure 14.19 (A) Collapsible grille; (B) roller grille made up of rods, tubular spacers and connecting links, as shown in detail.

Roller grilles (B) are popular for shop windows. The grilles may be used inside but more often they are fixed outside to protect the whole shopfront. Counterbalancing springs on the roller allow grilles of up to about 2500 mm (8 ft 3 in.) wide to be pushed up by hand; up to about 5000 mm (16 ft 6 in.) a geared operation using an endless chain or cranked handle is necessary, for greater widths grilles need to be electrically operated.

The horizontal rods of the grille may be solid or tubular, of diameters about 8—16 mm ($^5/_{16}$—$^5/_8$ in.) with other members of size to suit. Rods are usually spaced 50—90 mm (2—3½ in.) apart; connecting links are spaced at 57—300 mm (2¼—12 in.) giving meshes of varying size for selection according to the protection required. Though the grilles are not strong enough to resist a determined attack they will keep vandals at bay and be an additional obstacle for the determined criminal to overcome — even if he pulls the grille out of the way with a vehicle he will still have to get through the window or shopfront, and the grille will not give concealment.

Roller shutters are an alternative protection for shopfronts, but if a thief gets through one he may be able to attack the shopfront unobserved. Houses left unoccupied for long periods, such as holiday

Figure 14.20 A fortunate coincidence of architectural fashion and security —
steeply sloping sills will not hold terrorist bombs.

homes, can have their security improved by roller shutters fitted to
the windows.

For protection against terrorist attack roller grilles, or pull-up
grilles of similar appearance, may be used inside windows to resist
the penetration of bombs and missiles thrown against the glass.
(See section 6.4 for blast resistance of laminated glass.) Window
sills should slope steeply so that they do not form a ledge on which
bombs can be left (fig. 14.20).

Chapter 15

Security Alarm Systems

15.1 Requirements

Notwithstanding that full consideration has been given to design for security, it may be felt necessary to install a security alarm system as a second line of defence. No relaxation of security design can be permitted as a result of this decision. Security alarm systems are neither foolproof nor burglar-proof. They suffer from the following drawbacks:

- Alarm signals are ignored by the general public.
- Numerous false alarms detract from the effectiveness of the systems in alerting neighbours and police — over 95% of alarm calls received by the police are false.
- Annoyance of neighbours by false alarm signals can lead to prosecution in Britain under the Control of Pollution Act 1974 for committing a nuisance by noise, or perhaps prosecution for a breach of the peace — irate neighbours have taken the law into their own hands and ripped off alarm boxes.
- Persons responsible for business or industrial premises may be got out of bed in the middle of the night to turn off the alarm. Householders may be brought back from holiday for the same purpose.
- With the growth in the use of alarm systems, and the display of fake alarm boxes to give the impression that the building is protected, burglars are not so easily frightened off as they used to be. If there is a system installed they may ignore the alarm long enough to get away with valuable items or they may stop the alarm sounding by attacking the control box.
- The system may deliberately be rendered inoperative prior to entry of the burglar. This might be an 'inside job' by a dishonest or disgruntled employee, or it may be carried out by a make-believe customer or a seemingly-casual visitor to the building.

This list of shortcomings suggests some of the requirements of a good system: the alarm signal must demand attention; high priority must be given to freedom from false alarms; the system must be resistant to being tampered with prior to a break-in — this is more important in business and industrial premises than in domestic buildings; the control box must be robust enough to withstand attack, or alternatively, be hidden from sight; the rest of the system must also be resistant to deactivation after entry has been accomplished.

Other requirements are: the obvious one — that the system should detect intruders; that it should not be liable to breakdown, but if it does it should be as far as possible 'fail safe'; that its initial cost and running costs should be commensurate with the degree of protection deemed necessary.

15.2 Kinds of protection

There are three principal kinds of place in a building to guard with an alarm system: points, space and perimeter. Typical points to guard are doors and windows in external walls, trapdoors in ceilings, internal doors and places where items of value are displayed or stored. A number of sensors will be required, mostly for doors and windows. Fortunately sensors for these are cheap and reliable, their consumption of current is low and they are the type of sensor least likely to be triggered off accidentally when there is no one on the premises to deal with the alarm.

Space guarding covering areas of rooms, corridors and courtyards requires fewer but more elaborate (and expensive) sensors. These are liable to be activated by birds, animals, paper blowing about, traffic passing and air movement. When used outside they may be rendered inoperative by the weather. Precautions against these factors increase the complexity of the devices. They are needed however for use alone, where point sensors would be impractical, and as a back-up — burglars may keep away from entry points and break in through the wall or roof.

Perimeter guarding may be achieved by fixing sensors to fences and walls to detect movement or vibration, or by an invisible fence of beams radiated from point to point designed so that the alarm is actuated when the beam is broken.

The extent and complexity of the system necessary for a particular building will depend on the degree of protection required as well as the form of the building. For a house that is not much different from its neighbours the simplest and cheapest reliable system may be effective. Make it obvious that the house is protected and we hope the burglar will pick an easier target, or better still be altogether deterred, though there are dangers in this notion, as we have seen. A system giving 90% protection that is used (perhaps the front entrance door is not guarded, though internal doors are) is better than one designed for 100% protection but not switched on because accidental alarms have become troublesome or using a pass key is a nuisance. On business and industrial premises there may be more valuable or more easily disposed of goods. The premises may be obviously unoccupied at night and the system is more liable to misuse and being tampered with, hence a more elaborate system is required.

BS 4737 (1977—9) *Specifications for Intruder Alarm Systems in Buildings* (currently under revision) defines the standard required of a contractor in the installation and maintenance of systems with audible alarms only and systems with remote signalling of alarms. The requirements for various detection devices are also given. An inspecting organisation, the National Supervisory Council for Intruder Alarms, maintains a roll of approved contractors and ensures that those on the roll work to the requirements of the British Standard.

15.3 A basic system

A simple system suitable for domestic use is shown in fig. 15.1. Magnetic reed switches are used as point-of-entry sensors. Both types of switches shown in the figure work in the same way; when the door or window is shut the reeds of the switch are within the magnetic field of a nearby magnet, they are attracted to each other and thus the switch is closed. The magnet, encapsulated in plastic like the switch, is fitted into the edge of the door or screwed to the face of the opening light of the window. As long as the door and

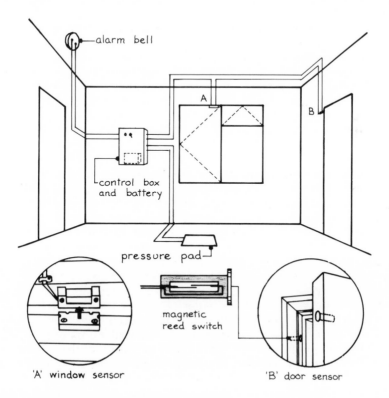

Figure 15.1 The basis of a simple alarm system.

window remain shut the switch is closed and current from the battery flows round the circuit, thus the circuit is of the *normally closed* type. It may be extended to include a number of doors and windows, often it is convenient to run the wire in a loop round the building, returning to the control box in the opposite direction from starting. Though the switches are generally used as shown, either type may be used on doors or windows; there are also types for special locations, such as sliding doors.

The pressure pad shown in fig. 15.1 operates in a *normally open* circuit. The pad usually consists of a PVC envelope containing two layers of metal foil separated by a layer of foamed plastics insulating material with holes in it at regular intervals. Pressure applied to the pad brings the two layers of foil into contact and the circuit is completed. BS 4737 requires an alarm condition to be produced by a force of 20—100N (4.5—22.5 lbf) applied normal to the surface by means of a solid disc 60 ± 5 mm (2.36 ± 0.2 in.) in diameter. The pads are concealed beneath carpets under windows, in hallways and on stairs, and in front of safes, trophy collections, etc.

When the normally closed circuit is broken or the normally open circuit is completed, a control diverts current to the alarm and latches it on so that it continues to sound even if the circuit is resorted — by the would-be burglar re-closing a door, perhaps. The alarm is switched off by means of a key inserted into a lock on the control box. The key is also used to arm the circuit; an indicator light on the control box warns if a door or window on the normally closed circuit has been left open. For the normally open circuit the indicator light will warn if anything is putting pressure on the pad.

In the simplest systems one door in the room or hallway where the control box is located has to be left unguarded to enable the keyholder to enter and leave without setting the alarm off, unless the control has a built-in entry/exit delay which allows a few seconds to get from the control to the door, or vice versa. In some systems sensors are fitted to interior doors only and windows are left unguarded on the assumption that a burglar cannot get far without opening a door.

Magnetic reed switches are extremely reliable. Unfortunately the same cannot be said for pressure pads: pets jump on them when left alone in the house; furniture inadvertently placed on a pad, while not heavy enough to give warning when the system is switched on, may settle slowly and cause a false alarm later. Also pads 'creep' and if they become buckled contact points may be brought together, perhaps causing an intermittent false alarm that will be difficult to locate: if carpets have to be taken up it may be necessary to employ expert carpet layers to replace them properly.

There are various ways of augmenting a simple system and making it more tamper resistant. A switch enables a guarded external door to be cut out of the circuit for the keyholder's exit and entrance. This switch may be incorporated in the lock or it can be separate and operated by its own key. The alarm (bell or siren, with or without flashing lights) can be made self actuating; it is provided with a separate power supply so that it will still sound if the wires between the control box and the alarm are cut. The battery in the control box may be supplanted by mains electricity, so avoiding troublesome battery replacement. However batteries last a long time with modern very low current systems, the trickle of current when the circuit is on guard is good for the battery, and providing the alarm does not make demands upon it the battery will last considerably longer than one that is not used at all — probably two or three years. Mains electricity is usually used through a rechargeable battery. Thus a standby is ready should there be a power failure. A self actuating alarm may also have a battery continuously charged from the control box.

A burglar may be able to locate a magnetic switch, say from outside a window, with a magnetic compass. If he then cuts a hole in the glass large enough to put his hand through, an external magnet can be used to hold the switch closed while the window is opened without actuating the alarm. Sensors that contain two reed switches each operated by magnetism of opposite polarity may be used to prevent such circumvention. Another way of cheating the alarm is to bridge the positive and negative wires coming from a sensor so that it is cut out of the circuit. Prior to entry of the building, the burglar or an accomplice removes the insulation from the wires at a point where they run closely together and brings the two wires into contact by twisting a short piece of wire between them. To prevent this a four wire system may be used. In this sensors contain two switches, one is closed when on guard the other open; there are two circuits, one normally closed the other normally open. To defeat the system the burglar must bridge two of the wires and cut one of the remaining two, but he does not know which to bridge and which to cut. Other alarm systems can have a 'day' or 'tamper' circuit running with the switched circuit. The additional circuit is kept active in the day and because it also presents the would-be tamperer with four wires, any attempt to interfere with the circuits is likely to set off the alarm. In business premises the door of the control box, and the casing of the alarm bell if it is accessible, should be made resistant to tampering by being guarded in the day as well as by night.

15.4 Wire barriers

The generally reliable, false-alarm free, normally closed circuit can be extended to protect walls, ceilings and doors by forming wire barriers, as shown in fig. 15.2. Wires are fixed about 100 mm (4 in.) apart and pulled taut so that they snap if an attacker attempts to pull them aside. The wires are covered, on walls and ceilings by dry linings, on doors by face panels. When used to protect windows (not just the glazing or the unauthorised opening of a light) the wires are guarded against accidental breakage by steel tubes. Wire barriers in partitions, doors and windows can be an effective way of protecting a storeroom or closet housing a safe or items of value.

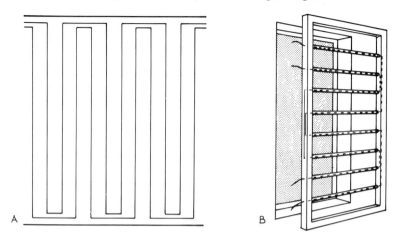

Figure 15.2 Wire barriers. Alarm sounds when wire is broken. (A) Wiring for walls, ceilings and doors; (B) wiring in tubes for window protection.

15.5 Guarding glazing

Simple methods of guarding glazed surfaces are shown in fig. 15.3. In one method a very thin strip of lead or aluminium stuck to the glass forms part of a normally closed circuit. To ensure that the metal strip breaks if the glass is broken the strip should not be placed too close to the perimeter of the pane, 50—100 mm (2—4 in.) is specified in BS 4737. Besides being visually obtrusive the strip is a hindrance to window cleaning. A less obtrusive thin wire embedded in the interlayer of laminated glass performs the same function as the metal strip with greater protection from accidental damage. Another form of laminated glass has an almost invisible electrically conductive layer within it. This layer is applied under heat to the inside of the glass on the side of the laminated panel open to attack. Along the top and bottom of the panel are electrodes and wired to

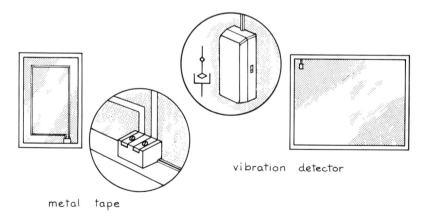

vibration detector

metal tape

Figure 15.3 Simple methods of guarding glazing.

the electrodes is an electronic detector. If the conductive layer is damaged because the glass to which it is bonded is cracked by an attack on it, the detector senses a change in the electrical resistance and actuates the alarm.

The principal advantage of the *vibration sensor* shown in fig. 15.3 is its cheapness. The sensor is stuck to the glass. Inside its plastics casing a pendulum makes contact when the glass is shaken sufficiently. It can be adjusted to ensure that an accidental knock should not cause the alarm to sound; usually it is adjusted until it fails to operate when the glass is struck by the fist. Nonetheless, false alarms are likely from sources of vibration: passing traffic, wind gusts, hail, doors slamming.

An *inertia sensor* is a more complex vibration sensor, more reliable and more expensive. It is used in a normally closed circuit in conjunction with an electronic analyser. The analyser triggers off the alarm when it senses that the vibrations are originated by deliberate action, for instance, when a specified number of impulses are received from the detector within a specified time, say ten in less than thirty seconds. The detectors can be used on walls and fences as well as on glazing, and a single analyser may monitor a number of detectors.

An *acoustic (or sound) sensor* consists of a microphone, filter and amplifier; it reacts to the sound of breaking glass. Sound waves picked up by the microphone produce electrical signals that are filtered so that only selected frequencies reach the amplifier which triggers off the alarm. Compact forms of this sensor can be stuck on to the glass that is to be guarded. Other forms operate up to 4—5 m (13—17 ft) away. Similar detectors that react to percussive sounds may be used for guarding walls and fences. Wherever used there is always the danger that an acoustic detector will pick up extraneous sounds composed of frequencies that trigger off the alarm.

15.6 Space sensors

A building with a large number of possible entry points will require a considerable amount of wiring to guard all points. Difficulties in ensuring that all point sensors are in the correct configuration and remain so when the system is on guard increase with the number of points. In such circumstances space sensors may be advantageous either as an alternative to point sensors or in combination with them. In any building, space sensors may be employed as a second line of defence. For existing buildings they have the advantage over point sensors in that they can be obtained as compact, self-contained units including an alarm and batteries. Even if connected to the mains for a power supply and to an alarm separate from the unit, they do not require the same amount of wiring as a system of point sensors.

There are four main types of space sensor:

- Ultrasonic
- Microwave
- Infra red
- Sound discriminating.

As their function is to sense movement, heat radiation or noise they are vulnerable to false alarms in many ways. Fluttering furnishings, the fall of a calendar held up by a drawing pin, water flowing in pipes, passing vehicles, draughts, lightning flashes and direct sunlight are just some of the things that may trigger off a false alarm. However a combination of the selection of the most suitable sensor for the purpose required, its siting with reference to happenings in the area, and the latest sophisticated circuitry to process the signal before actuating the alarm should reduce the false alarm rate to an acceptable level. As a further precaution a combined sensor (passive infra red and microwave, as described below) can be used. Both detection systems have to sense an intruder-type movement before the alarm is actuated.

Ultrasonic and microwave sensors

These sensors both operate on the same principle. Both are active sensors in that they transmit energy — sound waves of frequency above the level of hearing in one case and radio waves in the other — and both utilise the doppler effect (fig. 15.4). This is the apparent change in frequency due to relative motion between a source of waves and an observer. If the observer moves towards the source the distance between the waves he meets will be shortened, i.e. the frequency will be higher. If the observer moves away from the source the frequency will be lower. Thus the waves reflected off a moving object will differ in frequency from those reflected off a stationary object. When the receiver detects a frequency change it actuates an alarm.

Ultrasonic sensors have been prone to false alarms caused by

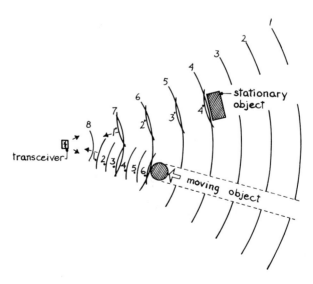

Figure 15.4 An impression of the doppler effect caused by a moving object — the principle of operation of space sensors.

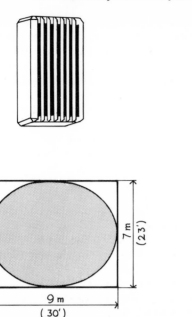

Figure 15.5 Ultrasonic and microwave sensors and typical floor coverage by ultrasonic sensor. The plastic cover on the microwave sensor is transparent to radio waves.

turbulent air from air conditioning and heaters, moving curtains, noise from water pipes and gas fires, telephone bells and vibration of the unit. With improvements in technnique they are now designed to eliminate, as much as possible, signals other than those produced by human movement. Microwave sensors analyse signals received to give a rough indication of an object's size and the distance it moves. The alarm is not actuated until an object reflecting a fairly large signal has moved about 1 m. This is to prevent false alarms being caused by animals or the movement of an inanimate object, a door perhaps. BS 4737 requires ultrasonic and microwave space sensors to at least detect the movement of a person weighing 40—80 kg (88—176 lb) moving through 2 m (6 ft 3 in.) at a speed of $0.3-0.6 \text{ ms}^{-1}$ ($^2/_3-1^1/_3$ m.p.h.).

The coverage of a typical ultrasonic sensor is sufficient to guard rooms up to about 7 x 9 m (23 x 30 ft) in area (fig. 15.5). This is capable of adjustment but the maximum coverage obtainable depends on the amount of reflection from walls and other surfaces. Reflections from such surfaces allow irregular shaped rooms to be effectively covered. Microwaves penetrate glazing, wood and plaster partitions and floors, and to some extent brick walls, but they are reflected by metal objects. It may be advantageous for microwaves to penetrate partitions made of 'soft' materials in, say, an office building or to penetrate packing cases in a warehouse but most undesirable that they should penetrate the floor or ceiling. To confine the transmission to a designated area and to allow adjustment, deflector plates that shape the beam can be fitted. Microwave sensors have a coverage greater than that of ultrasonic sensors, their range is typically up to 15 m (50 ft) with roughly the same sort of projection pattern as shown in fig. 15.5. In common with other types of space sensors, ultrasonic and microwave ones are provided with a light emitting diode (L.E.D.) which indicates the coverage without the alarm sounding when a 'walk test' is carried out.

Passive infra red sensors

'Active' space sensors such as ultrasonic and microwave sensors are those that continuously transmit energy when they are on guard. 'Passive' space sensors watch or listen for energy emanating from other sources, they receive but do not transmit. Infra red sensors for guarding space are of this type. They can be passive because a human body, and other matter, transfers heat by radiating infra red rays to anything that is at a lower temperature. These rays have a shorter wavelength than light but they travel in the same way. A typical sensor (fig. 15.6) has a series of parabolic-shaped mirrors that give it multiple viewing zones as shown in fig. 15.7. The infra red rays collected by the mirrors are focused on to a dual element pyroelectric sensor. Radiation from a moving object falls on first one element then the other; the sensor generates a signal when this happens. Local temperature changes should not give rise to false alarms as both pyroelectric sensors are equally affected. Stray air currents and sunlight are prevented from causing an actuating disturbance by the closing of doors and windows, thus effectively sealing the room. Rays generated by movement outside the room will not be detected because infra red radiation does not penetrate solids, including glass. Zones of cover may number fifteen or more and they may be arranged so that some are of short range, some medium and some long, with a range of up to 30 m (100 ft). The shorter zones are pointed downwards to prevent an intruder passing undetected beneath the pattern. The field of view given by the

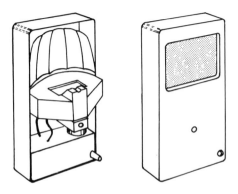

Figure 15.6 A passive infra red sensor (Ademco).

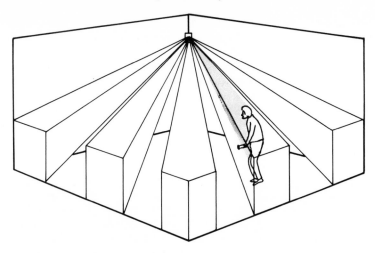

Figure 15.7 Zones of cover of a passive infra red sensor.

pattern can be arranged to be broad to cover a room or narrow for aisles and corridors. As an alternative to parabolic-shaped mirrors, detectors may be fitted with a Fresnel lens, i.e. a lens made up of a number of smaller lenses for the surveillance of zones.

Sound discriminating sensors
Providing force is used to attempt an entry, a sound discriminating sensor will sound the alarm while the intruder is still outside the building trying to get in. Also known as an *acoustic analysing sensor* it is a passive sensor, it listens to all sounds but reacts only to those made up of certain frequencies. It is more technically advanced than the acoustic sensor for glazing previously described, it reacts to sounds such as those made by the splintering of wood, metal striking metal, and breaking glass. Various features overcome the possibility of a single random sound creating a false alarm. Doors between rooms to be guarded must be left open to obtain maximum cover. The units are free-standing and transportable for use with or without a mains electricity supply. An optional remote sensor can be used in tandem with the main sensor, a remote alarm can also be used. Because they will ignore normal environmental sounds but react to noise generated by attack on the fabric of the building, sound discriminating sensors are useful for vandal protection.

15.7 Point to point sensors

These are active sensors: a beam of radiation is transmitted from one point and received at another. When the beam is broken the alarm sounds. Microwave or infra red radiation can be used. Infra red is usually preferred because the beam can be narrower than a microwave beam. Infra red also has the advantage, in some locations, that the beam can be made to change direction by reflecting it from a mirror in the same way that visible light can be reflected. A crystal of gallium arsenide, a semiconductor material, is used to generate the infra red radiation. An intruder may try to defeat the system by using another source to simulate the beam while he slips through it, an electric lamp (blackened so as not to give him away), glowing charcol in a can, even an ignited cigarette lighter, might be used — we have seen how hot bodies produce infra red radiation. To prevent the system being cheated in this way the transmitted radiation is sent in a series of rapid pulses; the alarm is triggered off if steady radiation is received in place of the pulsed radiation.

Generally infra red projectors, receivers and mirrors have similar cases. The unversed intruder will not recognise them for what they are because they have a light filter over the opening where the beam penetrates. Nonetheless the units are preferably hidden or disguised to prevent the more conversant intruder discerning the position of the beam and getting under or over it. To reduce false alarms BS 4737 requires that detectors do not operate for an interruption shorter than 2 milliseconds (ms). Outdoors, beams may be interrupted by birds, debris blowing about, falling leaves, even dust, so the delay is increased to 50 ms or more before the alarm is actuated.

Indoors, infra red, point to point systems are useful for the protection of long corridors, stairways, aisles in supermarkets, warehouses and factories, garage doors, doors of loading bays, ranges of roof lights and similar situations. Two applications are shown in fig. 15.8.

Outdoors, point to point sensors can form invisible fences, sense vehicles on approach roads and detect movement on roofs. If an outdoor infra red system is positioned where the beam is likely to be obstructed by a build-up of ice and snow on the units, then projectors and receivers fitted with automatic heaters must be used and the use of mirrors avoided.

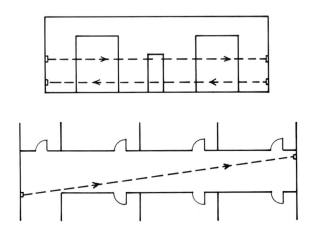

Figure 15.8 Typical indoor uses of infra red, point to point sensors — across doorways and along corridors.

Where cold weather is likely to prove troublesome the use of a microwave system as an alternative to an infra red system should be considered. Fog, rain and snow will attenuate microwaves but not to a notable effect. As with the microwave space detector, an analyser ensures that the alarm is actuated only when a fairly large object is detected. However the maximum range of microwave point to point systems is only about half that of infra red systems, these reach up to 300 m (1000 ft).

15.8 Perimeter protection

In addition to point to point sensors other ways of increasing the security of perimeters of sites are inertia and acoustic detectors, as described in section 15.5, secured to fencing.

Underground sensors for perimeter protection are also available. Flexible pressure-sensing tubes buried below ground convert a change of pressure to an electrical signal which is transmitted to an analyser. This differentiates between pressures from small animals, road traffic and intruders and actuates the alarm only when intruders are detected. Another system uses two or three coaxial cables, one of which acts as a radio transmitter while the other one or two act as receivers. Disturbance of the radio field by an intruder is detected by the receivers and the alarm generated.

Fibre optics have found an application in fencing, as shown in fig. 15.9. The cable used contains a thin (200 micron) continuous filament of glass which carries pulsed light signals round a system in a way analogous to the operation of a normally closed, electric alarm circuit. A break in the system causes an electronics unit to actuate the alarm and indicate the position of the break on a visual display unit. Repairs can be made without difficulty. Unlike electric wiring the fibre optic is unaffected by climatic, magnetic or electrostatic conditions, and it cannot be looped out.

The collapsible outrigger shown in D and E in fig. 15.9 may carry either plain or optical tape. The outrigger is hinged to the upright post and held in position by a hollow shear pin through which passes a fibre optic cable. When a weight of approximately 18 kg (40 lb) is imposed on the outrigger the pin and the fibre optic shear and the alarm is raised.

15.9 Other devices for alarm systems

Deliberately operated devices such as the so-called panic button or attack button may be included in an intruder alarm system (fig. 15.10). The circuits need to be arranged so that the device will be in the on-guard mode when the rest of the system is switched off, during the day for example. Also the device needs to be self-latching, e.g. the button stays in when pressed and can only be released with a key; alternatively there needs to be some form of electronic latching to keep the alarm sounding.

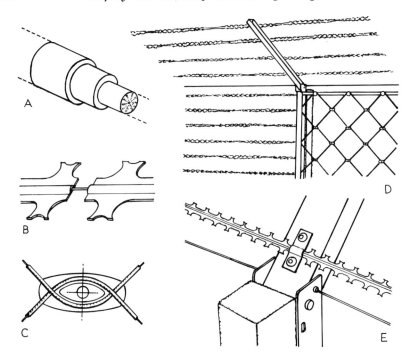

Figure 15.9 Fibre optic fencing (Pilkington). (A) Cable with resin coating and PVC protective sleeve; (B) barbed tape containing fibre optic; (C) link of mesh made from fibre optic cable; (D) alternative applications: barbed tape, and mesh; (E) outrigger, and cable that is sheared if outrigger drops.

Automatic light switches are primarily energy saving devices. By means of the doppler effect or sound discrimination a switch senses when a room is unoccupied and turns the light off. When a person enters the room it switches the light on again providing the ambient light is below a certain level. Thus a night-time intruder may be frightened off.

Wire-less systems commonly use tiny radio transmitters instead of wires to send signals from a sensor to an alarm; wire-less ultrasonic transmitters can be used with advantage providing the waves do not have to pass through walls. Though wiring for sensors is saved, a

Figure 15.10 Panic switches. (A) Button with key release; (B) foot rail for triggering alarm or surveillance camera.

wire-less system is costly not only because of the transmitters but also because each transmitter needs a power supply either from batteries or by plugging into the mains, and socket outlets need wiring. A variation of the radio wire-less system sends the signals from sensors through the electrical wiring of the building. One design uses a passive infra red sensor in combination with the transmitter and an alarm in combination with the receiver, each of the combined units being plugged into a socket outlet. This sort of system has obvious limitations but it is intended only for household use. It shares with the true wire-less system the advantage that the house-holder can take it with him when he moves.

Closed circuit television enables security guards to keep watch over indoor and outdoor areas and to visually investigate alarms without exposing themselves to risk of attack. Cameras can have remote-controlled pan-and-tilt heads and zoom lens, some cameras can see an intruder in almost complete darkness, other cameras take pictures with the aid of infra red spot lights that illuminate an intruder without him being aware of it.

15.10 Remote signalling

To ensure that action is taken when a sensor gives warning that an intruder has been detected the alarm may be signalled at a police station or a security company's central station. It can be arranged for the alarm on the premises to go off at the same time or after a predetermined delay to give the police a chance to apprehend the intruders before they are frightened off. With luck, and a fast response, the police will catch them as they run out. The different systems employed for remote signalling are shown in fig. 15.11.

A '999' dialler is the cheapest device to use for remote signalling. Its principal part is a small tape player. When an alarm is triggered the device automatically dials 999 (or other emergency number) over the normal telephone network and transmits a pre-recorded message. The 999 operator then informs the police. Care is needed to avoid failure of the system for the following reasons:

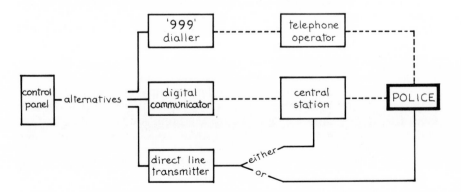

Figure 15.11 Remote signalling systems.

- The device does not work when required to do so. Maybe it has been installed for some years and no regular checks are carried out.
- The tape is unintelligible, either it has deteriorated or the message comes over in a strong ethnic or out-of-district accent.
- The telephone line is out of action. Possibly there is a fault on the line, possibly the intruder has engaged the line before breaking into the premises. This he can do by dialling the number and leaving the receiver off, most likely from a nearby call-box.

Thus regular maintenance and a clearly spoken message are essential requirements of a trouble-free system. The ploy of engaging the line can be thwarted by having the dialler connected to an 'out-going only' line, but an additional line increases the cost of the system.

A digital communicator when actuated dials the central station of a security company and transmits a coded signal. Separate connections of different sensor circuits to the communicator cause the code to indicate the kind of violation that has occurred, a door being opened for example. The advantage of this is that if it seems probable that a person has legitimately entered the premises but failed to carry out the correct procedures for deactivating the alarm the security company can telephone and ask the person for a prearranged response. In this way and by checks and maintenance the security company can filter out many of the troubles that arise from '999' diallers. Naturally the police prefer digital communicators.

A direct line transmitter is linked by direct telephone line either to a police station or to a commercial central station. In some areas a central station connection is the only option because the police authority will not permit connections directly into a police station. The reasons for this are the amount of surveillance required at the police terminal and the high proportion of false alarms: commonly in excess of 95% of all alarm calls. The connection is constantly monitored by pulses sent down the telephone line to ensure that contact is maintained. If the pulses are interrupted this is indicated at the terminal but the alarm is not given. Thus the reason for a line fault can be investigated without a full alert. If the system is being monitored by a central station the police will not be informed at this stage. The terminal indicates a full alarm only when the on-site equipment gives warning.

False alarms can be filtered out by a central station in the same way as with a digital communicator system. Opening and closing of the protected premises are kept under scrutiny and any departure from agreed times investigated. If a keyholder under duress departs from the agreed time for opening up the premises the central station will investigate, even though the correct opening procedures have been followed. A computer can be used to identify off-time signals and warn of signals not received, it can also be used to speed-up the response to alarms by retrieving information about the premises and the procedures the subscriber wishes to be followed.

A development of the direct line transmitter, called a *subscriber terminal unit*, does not require a private line. Information is transmitted over an ordinary line whether the telephone is on or off the hook and without disturbing a call that may be in progress. At the telephone exchange a scanner monitors the line and the subscriber unit. The scanner is connected to a computer which in turn is connected to a central station. These three items: subscriber unit, scanner and computer, are part of the telephone system, the security company pays a charge for the connection to the computer and passes this on to its customer in a periodical service charge for surveillance and maintenance of the sensor system. Heat and smoke sensors, panic buttons and medical alert buttons can be incorporated in the system.

15.11   Checklist

Guard:

		Reference to section
● *The perimeter*		
fences	with inertia or acoustic sensors or fibre optic barrier	15.5/8
inside a defined boundary	with microwave or infra red point-to-point, or underground sensors	15.6/8
● *Possible entry points into building*		
external doors		
windows that open		
trapdoors	with magnetic reed switches (or wire barriers to windows)	15.3/4
fixed glazing	with stuck-on metal strip, alarm glass, vibration, inertia or acoustic sensors	15.5
● *The structure*		
walls		
ceilings	with wire barriers, inertia or acoustic sensors	15.4/5
● *Selected points inside the building*		
internal doors	with magnetic reed switches	15.3
under windows		
passage ways		
stair treads		
in front of safes and collection cabinets	with pressure pads	15.3
● *Space*		
one room or other enclosed place	with ultrasonic or passive infra red sensors	15.6
one room or a number of rooms or other enclosed places separated by 'soft' partitions	with microwave or sound discriminating sensors	15.6

Suggestions for Further Reading

SAFETY

Acland, A.S. (1971) 'Safety in the home: metric edition.' *Department of the Environment Design Bulletin 13*. London: H.M.S.O.
Alessi, D, Brill, M. *et al*. (1978) *Home Safety Guidelines for Architects and Builders*. Washington D.C.: U.S. Department of Commerce.
Archea, J, Collins, B.L, Stahl, F.I. (1979) *Guidelines for Stair Safety*. Washington, D.C.: U.S. Department of Commerce.
Clark, A.J. and Webber, G.M.B. (1981) 'Accidents involving glass in domestic doors and windows: some implications for design.' *Building Research Establishment Information Paper IP 18/81*. Watford: B.R.E.

SECURITY

Anon. (1981) *The Essentials of Security Lighting*. London: The Electricity Council.
BS 0000 (In preparation) *Security of Buildings: Part 1: Dwellings*. London: British Standards Institution.
Building Research Establishment Digests and Information Papers.
Lyons, S.L. (1980) *Exterior Lighting for Industry and Security*. London: Applied Science Publishers.
Newman, O. (1972) *Defensible Space: People and Design in the Violent City*. New York: Macmillan. (1973) London: The Architectural Press.
Poyner, B. (1983) *Design Against Crime: Beyond Defensible Space*. London: Butterworths.
Stoik, J. ed. (1981) *Building Security*. Philadelphia: American Society for Testing and Materials.
Sykes, J. ed. (1979) *Designing Against Vandalism*. London: The Design Council.
Underwood, Grahame (1984) *The Security of Buildings*. London: The Architectural Press.

Safety Checklist

Hazard		Preventive measures and precautions	Reference to Section
	A	All users	
	C	Specially for child users	
	E	Specially for elderly users	

Hazard		Preventive measures and precautions	Reference to Section
Boundary walls, fences and gates			
Allows climbing	C	No hand or toe holds for climbing	7.1
	C	Minimum height 1200 mm (4 ft)	7.1
	C	On sloping ground, higher horizontal top edge extended 750 mm (2 ft 6 in.) beyond steps in wall or fence	7.1
Allows penetration	C	Impenetrable by 100 mm (4 in.) sphere	7.1
Spiked railings and other anti-scaling devices Spiked chains and other items capable of inflicting injury	A	Not used below 1800 mm (6 ft) from ground	7.1
Footpaths			
Slippery surface	A	Surface slip-resistant in wet conditions	10.3
Projecting windows Up-and-over doors Fuel supply inlets	A	Paths clear of buildings	8.2
High winds	E	Sheltering walls, handrails, avoidance of low-level 'trips'	8.2
Ramps			
Slope not noticeable	A	Change of level made distinct Length at least 900 mm (3 ft)	10.3
Elevated paths and areas			
Fall from height	A	Balustrade or barrier 1100 mm (3 ft 7½ in.) minimum height and as stairway balustrade, which see	10.4
Vehicle areas			
Moving vehicles	A	One-way routes	8.2
	A	Vehicle routes clearly distinguished from pedestrian areas	8.2
	CE	Vehicle routes separated from pedestrian areas	8.2

Hazard		Preventive measures and precautions A All users C Specially for child users E Specially for elderly users	Reference to Section
Blind exits	A	Barriers for pedestrians, slow bumps for vehicles	8.2
Vehicles out of control downhill	A	Earth bank or stout wall	8.2
Steps and stairways			
Change of level not noticeable	A	Replace by ramp, or explicitly indicate beginning and end of each flight, avoid single steps	11.2 11.3
		Provide landing at top and bottom (e.g. behind door)	
Imposition of awkward gait	A	Rise 100—180 mm (4—7 in.), going 280—355 mm (11—14 in.)	11.4
	A	In housing, if 32°—42° pitch unavoidable, $2r + g =$ 500—700 mm ($21\frac{5}{8}$—27½ in.), and g at least 240 mm (9½ in.)	11.4
	C	Preference for easy rise	11.4
	E	Preference for easy going	11.4
	A	Avoid triangular treads (winders)	11.4
	E	Avoid tapered treads	11.4
Distracting views or surroundings	A	Rounded nosings, lighting and other features to accentuate steps and handrails	11.3 9.3
		Avoid open risers	11.3
		Avoid abrupt change of view from stairway (orientation edge)	
Confusing pattern on tread surface	A	Avoid busy patterns and lines parallel with edge of tread	11.3
Water or ice on external steps	A	Treads perforated or well drained	11.5
Exacerbation of injury	A	Rounded nosings, and avoid sharp edges elsewhere	11.10
		Keep flights as short as practicable	
Lack of support on mis-step		Handrail	
	A	840—1000 mm (2 ft 9 in.—3 ft 3 in.) above pitch line if more than 600 mm (2 ft) total rise, each side if stairway over 1000 mm (3 ft 3 in.) wide, 1800 mm (6 ft) apart if wider	11.8
	C	600 mm (2 ft) above pitch line, as above	11.8
	E	for any rise, each side and not exceeding 1000 mm (3 ft 3 in.) apart	11.8
	A	rounded for comfortable grip and safe, emergency grip, diameter 45—50 mm (1¾—2 in.)	11.8
	A	clear of wall 65 mm (2⅝ in.) minimum	11.8
	A	uninterrupted in length	11.8
Start or finish not signalled	A	Handrails continued horizontally at top and bottom of stairway	11.8
Balance upset by clothing catching	A	Handrails returned or otherwise safely finished at ends	11.8
Elevated areas		Balustrade	
	A	900 mm (3 ft) minimum height in dwellings	11.8

Hazard		Preventive measures and precautions A All users C Specially for child users E Specially for elderly users	Reference to Section
	A	1100 mm (3 ft 7½ in.) minimum other buildings	11.8
	A	impenetrable by 300 mm (1 ft) diameter sphere	11.8
	C	impenetrable by 100 mm (4 in.) diameter sphere	11.8
	C	no hand or toe holds for climbing	11.8
Pedestrian area extends under sloping soffit	A	Guard where less than 2000 mm (6 ft 7 in.) headroom	11.6
Low soffit or edge of landing over steps	A	Minimum headroom 2000 mm (6 ft 7 in.)	4.6
Balustrade capping on which objects may be placed (e.g. glasses from bar) and fall from height	A	Avoid level surface	—
Slippery steps	A	Avoid polished wood, polished stone and smooth ceramic tiles Provide mat wells within buildings for removing moisture from users' feet	11.7
Foot-locks on tread surface	A	Avoid rubber matting and other surfaces where friction is too great to allow pivoting of foot	11.7
Glazing below 1500 mm (5 ft) above steps or landings	A	Safety glazing	14.1—2
High level glazing above steps or landings	A	Avoid unless provision made for safe cleaning and maintenance	11.9
Shadows cast on to steps	A	Avoid shadows parallel with treads	9.3
Escalators			
Fall on boarding or leaving	A	Travel on flat steps 0.8—1.2 m (2 ft 8 in.—4 ft) Landings 2—2.5 m (6 ft 7 in.—8 ft 3 in.) deep	11.11
Entrapment	A	Entry restrictors and deflector devices	11.11
All accidents	A	Emergency stop switch in prominent position Notice on proper use of escalator	11.11
Walls			
Rough textured surfaces Sharp arrises	C	Avoid	10.2
Dazzle from reflecting-glass cladding	A	No reflection of sun visible from road	10.2
Floors			
Slippery surface	A	Finish appropriate to probable footwear of users Finished selected for wet conditions where these likely No abrupt change of traction with adjacent surfaces	10.3

Hazard	Preventive measures and precautions A All users C Specially for child users E Specially for elderly users	Reference to Section
Roofs		
Maintenance work	A Fixed ladders with safety cage	10.4
	A Provision for resting and tying portable ladders	10.4
	A Eyebolts	14.4
	A Walkways and guard rails	10.4
Roofs with access and balconies		
Fall from height	A Balustrade 1100 mm (3 ft 7½ in.) minimum height and as stairway balustrade, which see	10.4
Doors		
Annealed glass	A Safety glazing below 1500 mm (5 ft) from floor	12.2
Swing door	Vision panel	
	A at eye level	12.3
	C to within 750 mm (2 ft 6 in.) from floor	12.3
	C Eliminate finger trap at hanging edge or provide finger guard	12.3
	E Lightweight construction or electro-mechanically operated by user floor springs	12.3
Door opening into area where users of furniture fitments or cookers will stand	A Re-locate or use sliding door	12.4
Doors close together on adjoining walls		
Cupboard door over worktop		
Open door projects into circulating area or escape route	A Recess doorway	12.4
Direct access to roadway	A Provide barrier	12.4
Up-and-over door	A Keep paths and steps away	12.4
Stairway nearby	Distance of doorway or swing-path of door from top or bottom of stairway at least	
	A 800 mm (2 ft 7½ in.) in domestic building	12.4
	A 1000 mm (3 ft 3½ in.) in other buildings	
Cupboard door over worktop	A Use sliding door	12.4
Lever handle	A Avoid type that hooks into clothing	12.5
Knob handle	A 75 mm (3 in.) clear of door edge	12.5
Windows		
Annealed glass	A Safety glazing within band 300—800 mm (1 ft—2 ft 7½ in.) from floor	14.2
Opening lights projecting into traffic route	A Avoid	8.2 14.2
Inside of glazing out of reach for non-professional cleaning	Keep within A 2100 mm (6 ft 11 in.) from floor	4.5

Hazard	Preventive measures and precautions A All users C Specially for child users E Specially for elderly users		Reference to Section
Fastener out of reach	E	1800 mm (5 ft 11 in.) from floor Maximum height from floor	4.5
	A	1900 mm (6 ft 3 in.)	4.5
	E	1570 mm (5 ft 2 in.) Maximum height with 600 mm (2 ft) obstructions in front	4.5
	A	1720 mm (5 ft 8 in.)	4.5
	E	1330 mm (4 ft 4 in.)	4.5
Openable window below 1350 mm (4 ft 5 in.)	C	Restrict opening to 100 mm (4 in.) maximum at bottom or side	14.3
		No window ledge that can act as a climbing aid	14.3
Cleaning outside of glazing	A	Provision for ladders for professional cleaning up to 9 m (29 ft 6 in.) from ground Cleaning from inside	14.4
	A	cleaner able to stand 1120 mm (3 ft 8 in.) below opening	14.4
	A	all parts within reach, 560 mm (1 ft 10 in.) sideways, 510 mm (1ft 8 in.) upwards, 610 mm (2 ft) downwards	14.4
	A	reliable reconnection of means of restricting opening Cleaning from walkways	14.3
	A	635 mm (2 ft 1 in.) minimum width	14.4
	A	balustrade	14.4
	A	safe access Cleaning (professionals only) from outside using walkways, working platform or travelling ladders	14.4
	A	provide lifeline attachments	14.4
	A	handholds	14.4
	A	safe access	14.4
Circulating areas Collisions with other people and equipment	A	Keep traffic routes direct	8.2
Pipe runs and service ducts Infection	A	No runs from non-hygienic areas through other parts of building	8.5
Workrooms Through traffic	A	Avoid	8.2
Kitchens Bad ergonomic relationship between activities, space and appliances	A	Provide sufficient space for appliances	8.3
	A	Give priority in planning to (1) sink, (2) cooker or hob, (3) mix, (4) other, (5) separate oven and freezer	8.3
Hobs exposed to children	C	Worktop at side, and at back for island hob unit	8.3

Hazard	Preventive measures and precautions A All users C Specially for child users E Specially for elderly users		Reference to Section
Items falling on to hob	A	No cupboard or shelves above cookers	8.3
Worktops on different levels	A	All at same level	8.3
Clutter		Provide sufficient storage and worktop space	8.3
Bathrooms			
Slipping in bath	A	Grab bars	8.4
Child climbing on appliances	C	Avoid climbing paths Light switch on wall outside bathroom	8.4
Electrocution	A	Safe heating	8.4
Infection	A	Washbasin with each WC	5.7
Lighting			
Glare from window	A	Light-coloured splayed reveals, baffles, tinted glass, another window at right-angles	9.2
Glare from luminaire	A	Relocate	9.2
Stairways and steps not clearly defined	A	Illuminate whole of tread with high-light on nosings Minimum of two lamps in case one fails	9.3
Electrical power outlets			
Out of comfort range	E	1000 mm (3 ft 3½ in.) from floor	9.6
Flex trailing	A	Adequate outlets to provide for alterations, none positioned above cooker	9.6
General			
Air pollutants	A	Avoid contaminated material Provide adequate ventilation	5.2—7
Water pollutants	A	Avoid lead in pipes and paint Keep system clean and treat circulating water	5.7

Security Checklist

(see section 15.11 for security alarms checklist.)

Vulnerability of feature	Preventive measures and precautions		Reference to Section
	V	Vandal resistance	
	B	Basic security	
	S	Strong security	
	M	Maximum security	

Boundary walls, fences and gates

Allows climbing	B	Maximum mesh size 50 mm (2 in.)	7.3
	B	Bracing rail instead of strut to straining post	7.3
	B	1.6 m (5 ft 3 in.) minimum height, 1.8 (6 ft) preferred	7.3
	S	2.1 m (7 ft) in height with anti-scaling precautions at top	7.3
	M	2.4 (8 ft) as above	7.3
	S	Change in direction not less than 130° in chain link or other mesh fence	7.3
Allows burrowing	S	Sill at least 225 mm (9 in.) below ground level, fencing anchored down if necessary	7.3
Gate fastenings can be forced	B	Stout locking bar and hinges	7.3—13.6
	B	Good quality lock	13.6/8
	B	Not possible to lift down-bolt out of ground when double gates closed	7.4
	S	Lock of appropriate grade	13.6—8
Can be penetrated by vehicles	S	Locking post on gates	7.6
	M	Trench or stout kerb 400 mm (1 ft 4 in.) high	7.3
	M	Rising road barriers	7.7
Subject to graffiti and impact damage	V	See Walls	

The site

Buildings hidden from casual surveillance	B	In view from roads	8.6
	B	Avoid nooks and crannies	8.6
	B	Not hidden by trees	8.6
	B	Security lighting provided	9.7

Vulnerability of feature	Preventive measures and precautions V Vandal resistance B Basic security S Strong security M Maximum security		Reference to Section
Footpaths			
Wrongdoers able to approach building without detection	B	Paths well away from buildings	8.6
	M	Controlled turnstile access	7.6/13.11
Attacks on pedestrians	B	Avoid sunken paths, tunnels and blind spots	8.6
Vehicle areas			
Wrongdoers able to enter and leave without detection	B	Roadways and car parks clear of buildings	8.6
	S	Driver operated barriers or supervised barriers	7.6
	M	Inspection of vehicles	7.6
Walls			
Graffiti		Surfaces	
	VB	hard	10.5
	VB	multi-coloured and patterned	10.5
	VB	textured	10.5
	VB	non-absorbent	10.5
Removal of copings	VB	Clip-copings	10.5
	VB	Coping imposing minimum load of 1.5 kN m^{-2}, or anchored	10.5
	VB	Copings keyed together	10.5
Impact	VB	Cladding with impact resistance of 6 Nm (4.4 ft lbf) or 10 Nm (7.4 ft lbf) according to location	10.5
Removal of cladding	VB	Avoid tile hanging and thin materials	10.5
Penetration	S	Avoid tile hanging	10.5
	S	Expanded metal or plywood under plasterboard linings of timber framed party walls	10.5
	S	As above under siding	10.5
	S	Separately fixed linings when metal sheeting used	10.5
	S	Stout kerb or limited space to prevent ramming	8.6
	M	Construction equal to 225 mm (9 in.) r.c. concrete for resistance to break-through	8.10
Openings allowing entry	B	No clear space greater than 0.03 m^2 (0.33 ft^2)	4.8
Floors			
Access to space under	B	Avoid	10.6
	S	Limit accessibility to removable panels	10.6
Roofs			
Access to flat roof	B	Avoid climbing routes	8.7
	B	Anti-climb guards or square-sectioned pipes tight to wall	10.8
Penetration	B	No vent or other opening greater than 0.03 m^2 (0.33 ft^2)	4.8
	B	Domelight resistant to removal	10.8

Vulnerability of feature	Preventive measures and precautions		Reference to Section
	V	Vandal resistance	
	B	Basic security	
	S	Strong security	
	M	Maximum security	
	S	Curbs resistant to prising up	10.8
	S	Burglar bars or polycarbonate laylight under roof lights	10.8
Ceilings			
Within reach	VB	Avoid soft materials	10.7
Access to space above	B	Tiles not removable	10.7
	B	Not continued over partitions	10.7
Balconies			
Access	B	Avoid climbing routes	8.7
	B	Separate balconies to each occupancy or divided by non-climbable screens	10.8
	S	Cage-like guarding	10.8
Doors			
Swinging on	VB	Three hinges and anchor hinge	12.10
	VB	Concealed door closer	12.10
Abuse of furniture	VB	Knob handles	12.10
	VB	Avoid handles at knee height	12.7
Unauthorised use of entrance	B	Can be seen from office, etc	8.8
Emergency exits	B	In plain view outside	8.8
	B	Surveillance possible from inside	8.8
	B	Single leaf or mullion if possible	13.5
Concealed by storm door	B	Fully (safety) glazed storm door	12.6
Penetration	B	Solid core or equal	12.8
	S	Plywood core	12.8
	M	Plywood core with steel plate	12.8
Glazing		See Windows	
Door frames			
Distortion	B	Stout construction	12.9
	S	Hardwood or steel (backed)	12.9
Fastenings	B	Fixings not more than 450 mm (18 in.) apart	12.9
	S	Steel rods to supplement fixing by screws	12.9
Locks and fasteners			
False key or picking	B	1000 differs	13.3/8
	M	1000+ differs, high resistance to picking	13.3
	B	Master key systems must not lower standard of resistance to picking	13.9
	B	Easily changed lock combination if large turnover of key holders	13.9
	B	Internally operated bolts on all but final exit door	
Tampering with fixing screws	VB	Backplate fixing or vandal resistant screws	13.7/3.6
Forcing lock	B	Withstand 13 000 N (3035 lbf) perpendicular to face	13.8

Vulnerability of feature	Preventive measures and precautions		Reference to Section
	V	Vandal resistance	
	B	Basic security	
	S	Strong security	
	M	Maximum security	
	B	Withstand 1200 N (270 lbf) against end of bolt	13.8
	S	Meet all BS 3621 requirements	13.3/8
	B	Use locking knob set (key-in-knob) only where forced entry would attract attention	13.3
Forcing door	B	Supplementary lock or multi-point locking on sliding patio doors	13.7—8
	B	Door limiter	13.7
	B	Concealed or key operated bolts on glazed double doors	13.4
	S	Strengthened staple	13.7
	S	Hanging edge fitted with dog bolts	13.4
	M	Multi-point locking	13.4
Turning key from letter plate	B	Letter plate at least 400 mm (16 in.) from lock	13.7
Windows			
Glazing	VB	Limit size of panes of annealed glass and make replacement easy, providing security not affected	—
	VB	No annealed glass at kicking height, near 'games' areas or unsupervised areas	6.1
	VB	Laminated glass to maintain security when fractured	6.4
	VS	Removable grilles for night-time use	14.8
	VS	Plastics glazing sheet materials	6.4
	B	Panes and louvres not easily removable from outside	14.6
	S	Louvres individually framed	14.6
	S	Cover with bars or grilles	14.8
		Use bandit-resistant glazing	6.4
	M	Use bullet-resistant glazing	6.4
Opening lights	VB	Strong frame and sashes and strong well fixed hinges to resist swinging on	—
	B	Too small for entry, 0.03 m^2 (0.33 ft^2) maximum	4.8
	B	Hinges, and nearby ventilators and fans, not removable from outside	14.6
	B	Window locks	14.7
	B	Avoid climbing routes to upper floor windows	8.7
	S	Provide bars or grilles	14.8
	M	No windows at ground floor level	3.5
Sanitary installations and associated finishings			
Items pulled away from fixings	VB	Robust construction and secure fixing	8.9

Vulnerability of feature	Preventive measures and precautions V Vandal resistance B Basic security S Strong security M Maximum security		Reference to Section
	VS	Concealed services	8.9
Can be unscrewed	VB	Tamper-resistant screws	3.6
	VB	Avoid bottle traps and other fittings that unscrew	8.9
Can be swung on, stood on, sat on	VB	Steel pipes	8.9
	VB	Strong fittings	8.9
	VM	Vandal-resistant fittings with button or foot operated valves and without waste plugs or WC seats	8.9
	VB	Washbasins not under windows nor adjacent to WC cubicles	8.9
Impact	VS	Vandal-resistant wall tiling	8.9
	VS	Stainless steel mirrors	8.9
Blockage	VB	Floor draining to gully	8.9
	VB	Rodding eyes and straight runs to inspection chambers	8.9
Cash offices			
Attack	B	Well inside building, no direct route from entrance, controlled approach for vehicles	8.7
	B	Airlock entrance	8.11
	B	Construction as for storerooms secure rooms and strongrooms	8.10
Payments counters			
Attack	B	Screens 2 m (6 ft 3 in.) high with 6.8 mm laminated glass	6.4/8.11
	S	Screen to ceiling with bandit-resistant glazing, 25 mm (1 in.) plywood counter top and front	6.4/8.11
	M	Screen with bullet-resistant glazing and bullet-resistant counter	6.4/8.11
Storerooms, secure rooms and strongrooms			
Attack	B	Block partitions, no access to ceiling, solid door, window lock	8.10
	S	100 mm (4 in.) brick reinforced partitions, plywood solid core door, barred windows	8.10
	M	Independent, fire resistant structure, 225 mm (9 in.) r.c. walls, no windows, steel door	8.10

Index

Included in the index are items in the safety checklist on pages 243 to 248 and in the security checklist on pages 249 to 253. These checklists may be used as a quick reference which will lead the reader on to sections giving details when amplification is desired.

Page numbers are in ordinary type for safety entries, in italics for security entries.